ALLE ZEIT WACH
SJ
1842

W. Braune O. Fischer

On the Centre of Gravity of the Human Body

as Related to the Equipment of the German Infantry Soldier

Translators:

P. G. J. Maquet R. Furlong

With 33 Figures and 51 Tables

Springer-Verlag
Berlin Heidelberg New York Tokyo 1985

Wilhelm Braune † Otto Fischer †

Translators:

Dr. Paul G. J. Maquet
25, Thier Bosset, B-4070 Aywaille, Belgium

Ronald Furlong, M.B., B.S., F.R.C.S.
149, Harley Street, GB-London W1N 2 DE, United Kingdom

Title of the original German edition: Über den Schwerpunkt des menschlichen Körpers mit Rücksicht auf die Ausrüstung des deutschen Infanteristen.
Published by S. Hirzel, 1889.

ISBN-13:978-3-642-69613-8 e-ISBN-13:978-3-642-69611-4
DOI: 10.1007/978-3-642-69611-4

Library of Congress Catologing in Publication Data. Braune, W. (Wilhelm) On the centre of gravity of the human body as related to the equipment of the German infantry soldier. Translation of: Über den Schwerpunkt des menschlichen Körpers mit Rücksicht auf die Ausrüstung des deutschen Infanteristen. Includes Index. 1. Posture. 2. Center of mass. 3. Body weight. 4. Soldiers–Germany. I. Fischer, O. (Otto). II. Title.
QP303.B7413 1985 612'.014 84-14132

Softcover reprint of the hardcover 1st edition 1985

2124/3140-543210

Foreword

It is matter of some surprise that this treasury of information concerning the centres of gravity of various parts of the human body has remained hidden not only from the English speaking world for obvious reasons, but also to some degree from the Germans themselves. What is less surprising is that this work is a monument to the renouned German thoroughness as demonstrated by the relentless pursuit of data and the meticulous accuracy of the conclusions.

However, these scientific investigations carried out in Leipzig must be viewed in the intellectual climate of the time. In the latter half of the nineteenth century, and later, Germany underwent an intellectual explosion. It was the age of Rhumkorff, Max Planck, the Weber brothers and Göttingen University. It was said that science came to life during discussions on the train between Göttingen and Berlin. So the scene was set and Braune and Fischer made good use of it and fulfilled their role as members of the Royal Scientific Society of Saxony.

When Pauwels (1935) analysed the static and dynamic forces exerted on the hip joint when standing and when walking, he needed data concerning the centres of gravity of the human body and of its different segments. I was faced with the same prerequisite when I studied the forces acting on the knee (1976). Both Pauwels and myself found all the data we needed, readily available, in the works of Braune and Fischer. I was amazed by the quality of the researches of these authors and, above all, by their astonishing accuracy. Unfortunately they had written in professoral German, using interminable and complicated sentences typical of the turn of the century. I thought that their results were still valid and could be of some use to many researchers. I asked my excellent friend, Mr. Ronald Furlong, F.R.C.S., for his enlightening help and together we spent months translating their works "On the Human Centre of Gravity" and "The Human Gait" into palatable English. We hope that our endeavour may be of some help to the scientific community.

The publishers are to be congratulated for their foresight in making these volumes available to a wider public than for which they were originally intended. In particular, Mr. Forster, copy editor, spent much time and labour in trying to improve our initial text which in some places was a little more than a literal translation. We are grateful for his endeavour.

Aywaille, October 1984 Paul G. J. Maquet

Table of Contents

Introduction

The effect of gravity, is such as to exert a force on each point of the mass of a body. Its line of action is the line connecting any one point of the mass with the centre of the earth. Its magnitude equals the weight of the mass imagined as concentrated at this point of the mass. As a result of the distance from the centre of the earth, all the lines of action of the forces of gravity are parallel. There is a resultant of the forces of gravity. For a certain position of a body, this resultant corresponds to a straight line inside the body. This straight line remains unchanged if the body is displaced a short distance in space, parallel to its long axis. It changes its position in the body, however, as soon as the long axis is turned. The straight line in the new position intersects the previous one at a particular point. The intersection point thus can be considered as the general point of application of all the resultants of the forces of gravity for any position of the body. This point is the centre of gravity of the body. The magnitude of the resultant is the weight of the body since the resultant equals the sum of the different forces of gravity. Consequently, we can imagine the weights of the different parts of the body as united at the centre of gravity of the body. The direction of the resultant is designated as the line of gravity. The line of gravity thus connects the centre of gravity and the centre of the earth. It does not change its absolute position in space as long as the centre of gravity is maintained, for example if the body is rotated any way about its centre of gravity. However, each rotation of the body changes the position of the centre of gravity in relation to the body. As long as the line of gravity intersects the support area of the body, the body stands; as soon as it comes to lie outside the support area, the body falls.

The support area is not necessarily the same as the surface through which the body is actually in contact with the ground. Both coincide only if the support area consists of one single surface, the outline of which is externally convex, as for example the base of a cone. If its outline presents indentations, i.e. if it is externally concave in places, the support surface comprises also the areas within the indentations, up to the double tangents that delineate the latter. This also applies if the indentations cross the whole surface, i.e. if the surface consists of two or more separate areas. The area between the two double tangents is part of the support surface.

When the human body stands on its two feet, the support surface involves the two soles and the area between the double tangents, which are drawn to the tips of the feet and to the heels. When erect, man can reduce his support area to a minimum by standing on the tip of one foot; he can increase it to a maximum by opening the legs as far as possible. This position is taken instinctively when fighting or when standing on a rolling boat deck. In this way, the line of gravity can undergo considerable oscillations without leaving the support area. If the support area is a point, the body is in equilibrium when the centre of gravity is on the plumb line of this point. The centre of gravity can take three positions relative to its supporting point.

1. The centre of gravity can coincide with the supporting point. In this instance, despite any rotation about the supporting point, the centre of gravity will keep its position in relation to the latter. In any position, the body remains in equilibrium. This equilibrium of the body is called neutral equilibrium.
2. The centre of gravity can lie above the supporting point. In this case, with each rotation of the body, the centre of gravity will leave the plumb line through the supporting point and take a lower position. An external force is, therefore, necessary to bring it back to its initial level. The centre of gravity will not come back by itself to its initial position. On the contrary, it will tend to fall further and further. As soon as it leaves the plumb line through the supporting point, the force of gravity exerts a turning moment. Consequently, the centre of gravity will keep falling until it has reached its lowest position. Such an equilibrium is called unstable equilibrium.
3. The centre of gravity can lie below the supporting point. In this case, a rotation of the body will raise it about the supporting point. Consequently, a turning moment will arise bringing the centre of gravity – always striving to fall – to its initial position, which is the lowest. Here the body must always return by itself to its position of equilibrium. This equilibrium is called stable equilibrium.

In unstable equilibrium, the centre of gravity lies at the highest position in relation to the supporting point; in stable equilibrium, it lies at the lowest position.

The lever arm that the force of gravity acquires when the centre of gravity rotates about the supporting point is smaller the closer the centre of gravity lies to the supporting point. It is zero when both coincide. This characterizes neutral equilibrium. Instability of the position in unstable equilibrium is greater the further the centre of gravity lies from the supporting point. This is experienced in everyday life: the closer the centre of gravity of the whole body lies to the ground the safer one is, because for the same displacement from the position of equilibrium the turning moment of the force of gravity is smaller the lower the centre of gravity lies. Consequently, short people are relatively more stable than tall people.

Statics and mechanics of the human body require precise knowledge of the position of the centre of gravity of the whole body and of its different parts as well as of the weights acting on these centres of gravity. Only such data can enable one to evaluate which force the muscles must exert in any movement of the body.

The human body is not a rigid mass. It has mobile limbs that change their position relative to each other all the time. They cannot even be considered as rigid in a mechanical sense since circulating fluids exist in the body. One could even wonder whether there can be such a thing as a centre of gravity of the human body. Using scales, Mosso[1] demonstrated that as a result of breathing and blood circulation the position of the total centre of gravity in the body continually changes. This change of position, however, is so small that it can be neglected for most of the problems of statics and mechanics of the human body. Furthermore, the changes in shape due to the soft tissues will exert some influence on the position of the centre of gravity. However, this is so small that, within certain limits, we can still consider the centre of gravity.

It is not the same, however, with the variable positions of the body segments. Because of the joints, the possibilities of displacement of these segments in relation to each other are such that a change in their position can displace the total centre of gravity of the body considerably. In the usual positions the total centre of gravity lies inside the body. However, the body can be curved to the extent that, as an arch, its centre of gravity comes to lie outside the body. Any determination of the centre of gravity thus is true only for a well-determined position of the body. Furthermore, determination of the centre of gravity can only be individual, to a certain point. This results from the differences in the distribution of mass in different individuals, depending on age, sex, etc..

Previously, mathematicians and anatomists have attempted to determine the position of the centre of gravity in the human body. However, their methods were insufficient because either they worked on soft cadavers or they reduced the latter to simple stereometric shapes. Therefore, it seems relevant to introduce a new method which avoids the aforementioned inconveniences by stiffening the body without adding any mass in such a way that it keeps its shape in any position.

The mathematician Borellus[2] determined the centre of gravity of the human body as follows. He laid a naked man on a board which was supported at its centre by a triangular wedge (op. cit. p. 167, propos. 134). There was equilibrium when the plumb line through the upper edge of the wedge lay between buttock and pubis.

Weber and Weber[3] attained greater accuracy. As Borellus had done, they tried to balance a body lying on a board upon a horizontal edge. To this end, they did not displace the board with the body but rather the body alone; the board had been balanced before the body came to lie on it. The measurement thus was completely independent of the board. Moreover, Weber and Weber did not try directly to determine the position of equilibrium of the body. This was impossible with such an unstable arrangement: the body was always more inclined to one side or the other. Weber and Weber determined the inclination of the board to one side and then repeated the measurement after the body had

---

[1] Mosso (1884) Application de la balance à l'étude de la circulation du sang chez l'homme. Archives italiennes de Biologie, tome V. Läscher, Turin, p. 131

[2] Joh. Alphonsus Borellus (1679) De motu animalium. Lugduni Batavorum

[3] Weber W, Weber E (1836) Mechanik der menschlichen Gehwerkzeuge. Göttingen, p. 114

been turned about on the board. Since one measurement indicated the centre of gravity as too high and the other gave it as correspondingly low the average of both gave the true position of the centre of gravity. After a series of measurements, all of which were independent of one another, they found that in a man 1669.2 mm tall, the distance of the centre of gravity from the vertex was 721.5 mm, from the heels 947.7 mm, and from the axis of rotation of the hips 87.7 mm. Then they measured the vertical distance of the sacral promontory from the axis of rotation of the hips in the skeleton of a man of the same height and found it to be 79.0 mm. Since this distance can be little different in men of the same height, this gave an opportunity for a fourth determination of the position of the centre of gravity to be made: its vertical distance from the sacral promontory was 8.7 mm. After a leg of the cadaver had been removed, the centre of gravity lay higher, approximately at the level of the umbilicus. After removal of the two legs, it lay still higher, at the level of the xiphoid process of the sternum.

According to Weber and Weber (op. cit., p. 119):

> One stands erect without any help from the muscles when knees and hips are fully extended beyond the position of the equilibrium. The centre of gravity of the trunk then lies above the ankle, the hip lies in front, and the knee is behind the plumb line through the centre of gravity. The compression exerted by the weight of the trunk tends further to extend the hips and the knees. This is hampered by the elasticity of their ligaments. The whole body down to the ankle is supported by the bones of the legs and their ligaments. It needs only to be balanced by the muscles above the ankle as one rigid mass.

Meyer thoroughly studied the position of the centre of gravity and above all the statics and mechanics of the human body[4].

Borellus and Weber and Weber made their measurements on a body lying supine. They assumed an absolute symmetry of the two sides of the body. The centre of gravity thus had to lie in the median plane. Borellus and Weber and Weber determined only its vertical distance from the vertex and from the feet. Meyer attempted to find the position of the centre of gravity in the antero-posterior direction. Therefore, he studied an erect position that he referred to and described as the military attitude. In his method, without any other change of position, the body was inclined forwards from the ankles until it almost tipped over about an axis drawn through the centre of the metatarsal heads of the great toes.

---

[4] Part of Meyer's works can be found in Müller's Archiv 1853 and 1854: das aufrechte Stehen, 1853, p. 9; das aufrechte Gehen, 1853, p. 365; die Mechanik des Kniegelenkes, 1853, p. 497; die Individualitäten des aufrechten Ganges, 1853, p. 548; über die normale Krümmung der Wirbelsäule, von Fr. Horner, mit einer Nachschrift von Meyer, 1854, p. 478. Part of Meyer's work can be found in the monographs: "Die wechselnde Lage des Schwerpunktes in dem menschlichen Körper, ein Beitrag zur plastischen Anatomie." Leipzig, Engelmann, 1863; "Statik und Mechanik des menschlichen Körpers". Engelmann, Leipzig, 1873

Similarly, extension at the ankle provoked a backwards inclination of the trunk and legs as a rigid unit until there was almost tipping over at the heel. This moved the centre of gravity over the whole supporting surface of the foot. The inclination to the horizontal was precisely measured at easily recognizable points of the leg in the initial position and in the two extreme positions. This gave the angular movement of the ankle. From this angular movement and from the horizontal displacement corresponding to the length of the foot, the position of the gravity line in the initial position could be determined.

In the erect position that Meyer referred to as "military" he found that the centre of gravity lay in the second sacral vertebra or just behind the vertebra in the sacral canal. He found the line of gravity to be 5 cm behind the axis of the hips and 3 cm in front of the axis of the malleoli[5]. Later he determined the centres of gravity of different parts of the body by reducing the head, trunk, and legs to simple mathematical bodies (ellipsoids and cones). According to him, the centre of gravity of the head lay a little in front of the occipito-atlantal joint. Meyer claims that the support of the head by the atlas strains the unstable equilibrium as little as the support of the body by the hip or the ankle joint[6]. The common centre of gravity of the head and trunk was found to be in the trunk by combining both individual centres of gravity. The common centre of gravity of the head and trunk, according to Meyer, lies a little in front of the middle of the tenth thoracic vertebra. He further divided the body. He considered the trunk as two ellipsoids, one corresponding to the thorax, the other to the abdomen. The centre of gravity of the whole leg was also determined – for a leg 720 mm long, it was found to lie 266 mm from the greater trochanter on the long axis of the leg. Meyer took the weights of the different segments from the text book of anatomy of Krause. Based on this material, he constructed a great many figures in different positions and showed how amazingly the position of the centre of gravity changes depending on the attitude of the trunk and limbs.

The next person to attempt a determination of the centre of gravity of the human body was Harless[7]. He repeated the experiment of Weber and Weber on a very well-built 24-year-old man. The subject's height was 1655 mm, the distance of the centre of gravity from mid-sole was 970 mm. For a standard height of 1000 this corresponds to a distance of 586.102 from the sole and 413.898 from the vertex. According to Weber and Weber, in a body 1669.2 mm tall, the distance of the centre of gravity from the heel is 947.7 mm. Thus for a height of 1000 the distance of the total centre of gravity from the vertex would be 432.24 and from the heel it would be 567.76. The difference between the data of Harless and those of Weber and Weber amounts to 1.8% of the overall height. Valentin (according to Harless) found a distance of 429.19 from the vertex for a

[5] Meyer (1873) Statik und Mechanik des menschlichen Knochengerüstes. Leipzig, p. 204

[6] Meyer, Horner, op. cit., p. 499

[7] Harless (1857) Die statischen Momente der menschlichen Gliedmaßen. Abhandl. der math.-physik. Klasse der Königl. Bayr. Akademie der Wissenschaften vol. 8, p. 71 ff, 1860, 1. Abth. München. 1857

height of 1000. According to Meyer the centre of gravity was 768 mm from the vertex in a body 1897 mm tall. For a height of 1000 this corresponds to a distance of 404.96 from the vertex and 595.14 from the sole. In this case the discrepancy in relation to the determination of Weber and Weber comprises 2.73% of the overall height.

For a height of 1000 the average distance of the centre of gravity from the vertex was 420.07 in four studies of different individuals, according to Harless. Harless attempted to obtain comparable values by reducing all the body heights to 1000 and by calculating the distance of the centre of gravity from the vertex and the heel on this basis. He did not take into account that the proportions of the limbs can be different in different individuals and that the level of the shoulders can vary considerably. For example, two bodies of the same height can have legs and trunks of different lengths; the distribution of the mass would then be different and this entails a difference in the position of the centre of gravity. A blacksmith with powerful arms set high must display a higher position of his centre of gravity than a tailor with weaker arms in a lower position, even if their heights are the same. Moreover, following on from the assumptions of Harless, it may happen that two bodies of the same height display the same distance between the centre of gravity and the vertex and however, great differences may exist in the position of the centre of gravity from an anatomical and mechanical point of view. For example, an individual may have very short legs and therefore his centre of gravity would lie high above the sacral promontory, whereas in another individual of the same height but with long legs it would lie below the sacral promontory, despite the same distance from the vertex. Thus, determining the position of the centre of gravity from its distance from the vertex has less value than determining it anatomically. This determination is, therefore, not improved by taking the arithmetical average of a number of measurements.

Harless attempted to determine the positions of the centres of gravity of the different parts of the body, as did Meyer and Horner, but in a more detailed and accurate manner. He used the oscillation method and thus obtained the height levels. From a consideration of their static moments he separated the limbs at the origin of the lever arms, i.e. at the axes of the joints. However, since he was operating on nonrigid material, he disarticulated the shoulders and the hips. After having repeated the Weber and Weber measurements on a living subject, he used the body of a beheaded man (named "Graf") to measure the positions of the centres of gravity of the different parts of the body (Table 1). To determine the centre of gravity of the trunk, he considered the latter as formed by two cones. To make sure that this assumption of the trunk as a system of cones did not entail too much inaccuracy, he calculated the weight of the trunk on this basis using the specific gravity of the whole body, assumed to be 1.066. For the whole trunk he estimated a weight of 28.5153 kg whereas direct weighing gave 29.608 kg. The specific gravity used, from the data of Harkness and Baumgärtner, was very high. According to Hermann, the specific gravity of normal cadavers is, on average, 0.9213[8]. Harless operated on a blood-

[8] Vierordt (1888) Tabellen. Jena

Table 1. Measurements made by Harless of different parts of the body

Part of body measured	
Height of body	172.685 cm
Height of foot	6 cm
Length of lower leg	42.9 cm
Length of thigh	44.9 cm
Distance of greater trochanter from apex of iliac crest	14 cm
Distance of ilium from top of acromion	39 cm
Height of head, from chin to vertex	21.2 cm
Anterior length of neck	4.7 cm
Length of foot	25.369 cm
Length of arm	86.6 cm
Length of upper arm	36.4 cm
Length of forearm	29.889 cm
Length of hand	20.314 cm
Total weight before execution	63 970 g
Weight of head[a]	4 555 g
Weight of trunk[a]	29 608 g
Weight of two arms[a]	7 540 g
Weight of two legs[a]	22 270 g
Average weight of hand	540 g
Weight of forearm	1 160 g
Weight of thigh	7 165 g
Weight of lower leg	2 800 g
Weight of foot	1 170 g

[a] The body was of a man who had been executed; it was bloodless.

less cadaver. Therefore, the difference between his calculation and the direct weighing of the trunk with its blood would have been much greater. In assuming an even distribution of mass in the trunk, considering the latter in terms of mathematical bodies would already give considerable errors of weight. However, mass is very unevenly distributed in the trunk because of the eccentric posterior situation of the heavy spine. This makes the determination of the centre of gravity by this mathematical method much more unreliable.

Direct measurements of the position of the centre of gravity of the limbs gave the following average distances:

1. 17.621 cm from the upper end of the 36.4-cm-long upper arm, 18.779 cm from its lower end
2. 13.122 cm from the upper end of the 29.889-cm-long forearm, 16.767 cm from its lower end
3. 9.623 cm from the upper end of the 20.314-cm-long hand, 10.691 cm from its lower end
4. 20.995 cm from the upper end of the 44.9-cm-long thigh, 23.905 cm from its lower end

5. 15.455 cm from the upper end of the 42.9-cm-long lower leg,
   27.445 cm from its lower end
6. 11.664 cm from the posterior end of the 25.369-cm-long foot,
   13.705 cm from its anterior end
7. 7.7 cm from the vertex of the head, 21.2 cm high,
   13.5 cm from the chin
8. The centre of gravity of the upper part of the trunk lay 23.465 cm from its lower limit, 17.53 cm from its upper limit

   The centre of gravity of the lower part of the trunk lay 5.8899 cm from its upper limit and 7.6101 cm from its lower limit.

Harless divided the trunk into an upper part down to the iliac crests and a lower part comprising the pelvis. Measurements and calculations were carried out to hundredths or even thousandths of a millimetre. This is as incomprehensible as it is useless.

Using the distances of the centres of gravity from the ends of the limbs of the executed man, the author calculated the position of the overall centre of gravity of the body. If the length of the body is reduced to 1000, the distance of the centre of gravity from the vertex is 413.65. Previously Harless had determined the centre of gravity of a 24-year-old well-built, living man directly according to the method of Weber and Weber. He found a distance of 413.898 from the vertex for a height of 1000. Both results coincide. However, this agreement is no proof of the accuracy of the direct or of the indirect measurement, for example when Harless considers the trunk as a system of geometric bodies, because the determinations relate to two completely different individuals. It must be mentioned that in the research of Weber and Weber, the distance of the centre of gravity from the vertex was 432.24 for a total height of 1000.

In a second contribution[9], Harless gives the sizes and weights of a second cadaver, that of a 29-year-old man named "Kefer" who had been beheaded; the cadaver was bloodless (Table 2).

Data concerning the position of the centres of gravity will be dealt later with our own findings.

Harless determined only the height of the positions of the centres of gravity; these positions in the two other directions thus remain to be determined. Such a direct determination of the centre of gravity of the whole system has as its control the calculation of the positions of this centre of gravity from the centres of gravity of the different body segments arrived at directly. Inversely it provides a control of or the experimental determination of the centres of gravity of these different segments. Neither Meyer nor Harless performed such a study; neither determined both experimentally *and* by calculation the same centre of gravity of the whole body or of a system of body segments in one and the same individual. Only an investigation of this kind can provide a reciprocal control of the two approaches. Harless compared the results of his calculation, which concerned only the heights, with the experimental findings. But these were measurements carried out on another individual.

Table 2. Data concerning the second cadaver measured by Harless

Part of the body measured	Right		Left
Weight (g)			
Whole cadaver		47 087	
Head		3 747	
Trunk (bloodless) without head and limbs		19 846.5	
Upper arm	1 484.5		1 411.3
Forearm	821.1		770.1
Hand	393.2		374
Thigh	5 947		5 827
Lower leg	2 242.6		2 252.4
Foot	982.2		988.2
Length, height (cm)			
Length of whole cadaver		167.85	
Height of head		20.2	
Length of trunk		57.5	
Length of thigh		42.3	
Length of lower leg		38.15	
Height of foot		9.7	
Length of foot		25	
Width of foot at heel		6.3	
Width of foot at metatarsal heads		9.9	
Length of upper arm	30.6		30
Length of forearm	26.4		26.1
Length of hand	18.5		18.9
Distance between 2 iliac crests		27.1	
Distance between centres of rotation of shoulders		29.9	
Distance between centres of rotation of hips		16.8	

[9] Harless, op. cit. p. 257

Determining the Position of the Centre of Gravity in the Cadaver

In our research on the position of the centre of gravity about to be described, we used rigidly frozen cadavers in order to avoid errors resulting from the softness of the material. Using an artificial freezing mixture we froze the cadavers so hard that they remained completely rigid until the end of the investigation, both as whole cadavers and as the segments into which they were later divided.

The method of determining exactly the centre of gravity of a frozen cadaver or of a part of it is described below. Repeated experiments have shown this method to be the most practical. It would be sufficient to hang the cadaver by a rope from two different points. Each time the cadaver was suspended it would oscillate until the centre of gravity came to lie vertically beneath the point of suspension, i.e. where the rope would continue through the body. The centre of gravity would lie at the intersection of the continuations of the two ropes. However, tracing the directions of the ropes through the body poses practical difficulties and hence must lead to inaccuracies. Therefore, this method was discarded.

We did not hang the cadaver from a rope but on an axis. This axis consisted of a steel rod as thin as possible but strong enough not to bend when supporting the cadaver. To determine the centre of gravity of the whole body, stronger steel rods were employed than those used to determine the centre of gravity of the limbs, and for the limbs the rods were stronger than for the lower leg or for the foot. For the latter, very thin steel needles were sufficient. By hanging the body or a segment of it from three different axes, it was possible to determine three different planes almost perpendicular to each other. The centre of gravity must lie in each of them and was thus at the intersection of the three planes. The rods were hardened and pointed so that they could be easily hammered in the chosen direction through all the tissues of the frozen cadaver, including the bones, with a heavy hammer. Since they were circular in cross section they formed an axis about which the cadaver easily rotated and oscillated until the centre of gravity came to lie vertically under the axis. Finding the position was very easy and reliable since we always introduced the axes as far as possible from the assumed position of the centre of gravity. Using two plumb lines attached to the axis, we determined each time the plane in which the centre of gravity lay and we marked on the surface of the cadaver the intersection of the latter by this plane. We accurately found the line of intersection even in the cases in which the plumb lines could not be hung in the immediate vicinity of the point of entry of the axis. To this end, the plumb lines were chosen long enough so as to be seen underneath the superimposed body segments.

Theoretically, the position of this system of planes in relation to the body is irrelevant. For practical reasons, however, we arranged for one plane to be almost perpendicular to the long axis of the body or of the body segment in question. This was achieved by first laying the body or the body segment with its long axis on a straight edge and balancing it by trial and error. We then introduced the first axis parallel to the straight edge, as close as possible to the upper surface of the body. Using the plumb lines, we delineated and marked this plane on the surface of the body. The body was then suspended in a second position. The second axis was introduced in a sagittal direction, as far as possible from the transverse marked plane. The second plane was then marked on the body. The intersection of these marks with the first ones was also marked. The body was suspended for the third time in the same manner. The third axis was introduced as coronal as possible, perpendicular to the second, and as far away from the transverse plane as possible. The points of intersection of the new plane with the transverse plane were marked on the surface of the body.

The cadaver was then sawn transversely. Two thin threads joining the intersections on the surface of the body were put on the flat hard surface thus obtained. These two threads marked the intersection of the transverse plane and the two others. The crossing of the threads gave the position of the centre of gravity. Since the second and third axes were perpendicular, these threads crossed each other almost perpendicularly. This made it easier to find the points of intersection.

We had at our disposal the cadavers of four normally built males (suicides). The first was used only to determine the overall centre of gravity since we were not allowed to saw it. In the second cadaver, the centre of gravity of the whole body and that of the trunk could not be determined with absolute reliability because during the measurement, parts of the trunk began to thaw and thus changed their position. Only the two last cadavers could be used for all the measurements.

Cadaver 1

The first cadaver was that of an 18-year-old, well-built, and muscular young man (height 169 cm), who had shot himself. It arrived at the department of anatomy fresh with pronounced postmortem rigidity. The abdomen was not swollen but rather sunken. We were not allowed to saw the corpse. Therefore, in this case, we could not carry out completely the aforementioned method. We had to restrict ourselves to suspending the cadaver in three positions as described above and to scratching the delineation of the three planes on the surface. After taking measurements from the bony prominences we transferred the intersections of the marks of the planes to the skeleton of a man of the same size and put the threads through this skeleton. The centre of gravity of the whole body thus arrived at, lay at the level of the lower edge of the second sacral vertebra, almost in the plane of entry to the pelvis, 0.5 cm to the right of the median plane, 4.5 cm below the sacral promontory, in the plane through the sacral promontory and the centres of the femoral heads. During freezing the cadaver lay supine on a horizontal board. The head was not raised and the limbs were symmetrically placed; the legs were extended, slightly externally rotated; the elbows were slightly flexed with the forearms half pronated.

Cadaver 2

The cadaver was that of a strong, muscular, 45-year-old man who had hanged himself. His abdomen was not protruding but rather sunken so that the inferior limit of the thorax was well marked. The length of the corpse was 170 cm, the weight was 75 100 g. It was frozen in the same position as cadaver 1.

After all the marks had been scratched onto the frozen cadaver as described above, we sawed off the head and the four limbs, which we put back into the freezing mixture to maintain their rigidity. The trunk alone was hung in three different ways and the marks for the determination of the centre of gravity were scratched onto its surface. This took time since our technique at this stage had not yet sufficiently developed to allow us to carry out the measurements quickly enough and with sufficient accuracy. During the measurements, the trunk began to thaw as mentioned above. The saws could no longer cut along the transverse planes. We could only pass thin steel needles through the cadaver at the appropriate points to mark the intersections of the planes and thus determine the overall centre of gravity and that of the trunk, as far as this was still possible.

We were only able to ascertain that the overall centre of gravity lay in the vicinity of, but below, the sacral promontory. The centre of gravity of the trunk could be determined with somewhat greater accuracy. It lay almost in the median plane at the lower edge of the first lumbar vertebra at the level of the intervertebral disc. The weight of the isolated trunk was 36 020 g. Except for these measurements, which were approximate, we were able to carry out all the other determinations just as well on this cadaver as on the subsequent ones, following the method described above.

The centre of gravity of the head lay exactly in the median plane on the clivus at the junction of the sphenoid and the occipital bone, below the dorsum sellae. It must be noted that the head had been cut along with part of the neck. The cut through the neck was dorsally oblique 5 cm above the manubrium sterni. Therefore, the centre of gravity we found was a little lower and more anterior than that of the head without the neck. The head with the neck weighed 5350 g.

The arms were cut along a sagittal plane upwards from the axillae in such a way that the saw passed almost through the centre of the head of the humerus. After being separated from the trunk, the arms were weighed. The right arm weighed 4950 g, the left 4790 g. The centre of gravity of the whole arm was determined after the planes had been marked. The forearm was then separated from the upper arm by a cut following as precisely as possible the axis of the humero-ulnar joint. It was seen later that the cut did not exactly coincide with this axis but was somewhat lower, such that in both cases the upper arm appeared a little too long in relation to the forearm. As it happened, the centre of gravity of the arm lay in this plane of section. The centre of gravity of the upper arm was then determined as well as that of the whole limb. Prior to this, the upper arm, forearm, and hand had been weighed. The centre of gravity of the forearm with the hand was determined but the planes were only temporarily marked. The hand was then separated transversely through the axis of flexion of the wrist, through the head of the capitate. It was thus possible to determine the weight and centre of gravity of the forearm with the hand, the forearm alone, and the hand alone.

The length of the right upper arm was measured from the centre of the head of the humerus to the surface of the cut, which in this cadaver passed almost through the joint space rather than along the axis of the humero-ulnar joint. The length was 31.7 cm and the weight 2580 g. The length of the right forearm without the hand to the surface of the cut, which went trought the head of the capitate, was 29.5 cm. The right forearm with the hand weighed 2370 g, the right forearm alone 1700 g, and the right hand alone 670 g. The left upper arm was also cut through the joint space. Its length was 31.5 cm, and its weight 2560 g. The length of the left forearm was 29.5 cm. The left forearm with the hand weighed 2230 g, the left forearm alone 1600 g, and the hand alone 620 g.

The centre of gravity of the right upper arm lay in the medullary cavity of the humerus 14.5 cm from the centre of the head of the humerus and 17.2 cm from the joint space of the humero-ulnar joint. The centre of gravity of the whole right arm lay in the cut through the elbow joint space; it was a little below the ulnar axis, in the lower end of the brachialis anticus, almost in the elbow crease, 1.5 cm in front of the bone between the origin of the flexors and extensors. The centre of gravity of the right forearm lay 12.5 cm below the humero-ulnar joint space and 17 cm above the head of the capitate, in the flexor muscles, 1 cm in front of the middle of the interosseous membrane. The centre of gravity of the right forearm with the hand lay 19 cm below the humero-ulnar joint space and 10.5 cm above the head of the capitate, in the flexor muscles, 0.5 cm in front of the insertion of the interosseous membrane on the radius. The centre of gravity of the right hand with partially flexed fingers lay 5.5 cm below the centre of the head of the capitate, in the palm, 1 cm in front of the bone.

The centre of gravity of the left upper arm lay 13.3 cm below the centre of the head of the humerus and 18.2 cm above the humero-ulnar joint space, in the medullary cavity of the humerus. The centre of gravity of the whole left arm was also located in the humero-ulnar joint, a little below the axis of the ulna in the middle of the elbow crease in the lower end of the brachialis anticus, 2 cm in front of the supratrochlear fossa. The centre of gravity of the left forearm lay 12.4 cm below the humero-ulnar joint and 17.1 cm above the head of the capitate, 1 cm in front of the middle of the interosseous membrane, in the flexor muscles. The centre of gravity of the left forearm with the hand lay 19 cm below the humero-ulnar joint space and 10.5 cm above the head of the capitate, 1 cm in front of the insertion of the interosseous membrane on the radius, in the flexor muscles. The centre of gravity of the left hand with the fingers partially flexed was exactly as in the opposite hand.

The lower limbs were separated by cutting through the hip joints. The cuts were made symmetrically through the femoral heads, as was later established. The centre of gravity of the whole leg was first determined and the intersecting planes temporarily marked on the surface. The thigh was then separated by a transverse cut through the axis of the knee joint. The centre of gravity of the thigh was then determined as well as that of the whole leg, which was also found to be located in the thigh. The centre of gravity of the lower leg with the foot was determined, the planes being marked only temporarily. The foot was separated from the lower leg transversely through the axis of the ankle joint. The centres of gravity were then established for the lower leg alone and the lower leg with the foot, the centre of gravity of the latter being located in the lower leg. Finally, we determined the centre of gravity of the foot as described above. At the same time, the segments were weighed.

The right leg weighed 12 120 g, the left 11 850 g[10]. The length of the right thigh was measured from the centre of the femoral head to the axis of the knee joint and was 44 cm; its weight was 7650 g. The right lower leg without the foot, from the axis of the knee to that of the ankle, measured 41.4 cm. The lower leg with the foot weighed 4470 g, the lower leg alone 3210 g, and the foot alone 1100 g. The right foot was 28.5 cm long and 7.8 cm high. The left thigh measured 43.3 cm; its weight was 7300 g. The length of the left lower leg was 41.8 cm. The left lower leg with the foot weighed 4500 g, the lower leg alone 3320 g, and the foot alone 1160 g. The left foot was 28.3 cm long and 7.5 cm high.

The centre of gravity of the right thigh lay 19 cm below the centre of the head of the femur and 25 cm above the axis of flexion of the knee, 1.5 cm dorsal and somewhat medial to the linea aspera. The centre of gravity of the right leg lay 39 cm below the centre of the head of the femur and 5 cm above the axis of flexion of the knee, in the middle of the posterior aspect of the bone. The centre of gravity of the right lower leg lay 17.4 cm below the axis of flexion of the knee and 24 cm above the axis of the ankle, 1 cm behind the middle of the interosseous membrane, exactly equidistant from the tibia and fibula. The centre of gravity of the right lower leg with the foot was found to be 24.6 cm below the axis of flexion of the knee and 16.8 cm above the axis of the ankle, directly behind the insertion of the interosseous membrane onto the tibia. The centre of gravity of the right foot lay 11.5 cm from the posterior limit of the foot and 17 cm from its tip. It lay thus 6.5 cm in front of the middle of the ankle joint, immediately behind the joint between the second and third cuneiform bones, on the anterior surface of the navicular bone.

The centre of gravity of the left thigh lay 19.3 cm below the centre of the head of the femur and 24 cm above the axis of the knee. The centre of gravity of the left leg lay 38.5 cm below the centre of the head of the femur and 4.8 cm above the axis of the flexion of the knee. The centre of gravity of the left lower leg lay 17.4 cm below the axis of flexion of the knee and 24.4 cm above that of the ankle. The centre of gravity of the left lower leg with the foot lay 25.5 cm

[10] Note, there is a discrepancy between the weight of the left leg as cited here and as given in Table 3. This discrepancy occurs in the original 1889 edition of this work.

Table 3. Data concerning cadaver 2

Part of body measured	Right		Left
Weight (g)			
Whole body		75 100	
Head		5 350	
Trunk without limbs		36 020	
Arm	4 950		4 790
Upper arm	2 580		2 560
Forearm with hand	2 370		2 230
Forearm without hand	1 700		1 600
Hand	670		620
Leg	12 120		11 890
Thigh	7 650		7 300
Lower leg with foot	4 470		4 500
Lower leg without foot	3 210		3 320
Foot	1 100		1 160
Length, height (cm)			
Length of whole cadaver		170	
Length of head[a]		21	
Length of thigh	44		43.3
Length of lower leg	41.4		41.8
Height of foot	7.8		7.5
Length of foot	28.5		28.3
Distance between the centres of the femoral heads		17	
Length of upper arm	31.7		31.5
Length of forearm without hand	29.5		29.5

[a] From chin to vertex in vertical projection

below the axis of the knee and 16.3 cm above that of the ankle. The centre of gravity of the left foot lay 12 cm from the posterior limit of the foot and 16.3 cm from its tip. The centres of gravity within the left leg were as in the right limb.

The measurements made on cadaver 2 are given in Table 3. The weights were measured directly. The sum of the individual weights may not reach the weight of the whole since some material was lost in the cutting. The masses are always considered from one articular axis (or centre of the joint) to the other, as long as the cuts through the joints coincided with the axes. Only in the elbow did the cut in both arms pass through the humero-ulnar joint space, as mentioned above.

Cadaver 3

This was the fresh cadaver of a normally built manual worker about 50 years old with strong muscles who had hanged himself. The fresh cadaver was frozen on its back as the others. When solidly frozen it was used to determine the centres of gravity as described above. It was 166 cm long and weighed 60 750 g.

The centre of gravity of the whole body lay in a transverse plane through the lower border of the first sacral vertebra, near the sacral promontory though

Table 4. Data concerning cadaver 3

Part of body measured	Right		Left
Weight (g)			
Whole body		60 750	
Head		4 040	
Trunk without limbs		28 850	
Arm	3 550		3 480
Upper arm	1 990		1 880
Forearm with hand	1 550		1 600
Forearm without hand	1 050		1 120
Hand	500		470
Leg	10 650		10 250
Thigh	6 690		6 220
Lower leg with foot	3 950		3 980
Lower leg without foot	2 870		2 880
Foot	1 060		1 090
Length, height (cm)			
Length of whole cadaver		166	
Length of head [a]		20.2	
Length of leg [b]	92.7		91.6
Length of thigh [b]	42.0		41.0
Length of lower leg [b]	43.0		42.9
Height of foot	7.7		7.7
Length of foot	26.5		26.9
Distance between the centres of the femoral heads		17.5	
Length of upper arm [b]	30.6		30.2
Length of forearm without hand [b]	26.3		27.1

[a] From chin to vertex in vertical projection
[b] The segments are always considered from one articular axis (or centre of joint) to the other

somewhat lower. It lay 0.2 cm to the right of the median plane and 4 cm in front of the upper transverse diameter of the sacrum. The centre of gravity of the trunk, without the head and arms, lay in the middle of the first lumbar vertebral body, 2 cm from the front, 1.4 cm from the back, and 0.3 cm to the right of the median plane. The centre of gravity of the head lay in the fossa behind the dorsum sellae, exactly in the median plane.

In this cadaver, the cuts passed exactly through the centres of the humeral heads and coincided with the axes of flexion of the humero-ulnar joint and of the wrist. See Table 4 for the measurements made on cadaver 3. Here also, the sum of the weights of the different segments is smaller than the total weight because the weights of the whole limbs and those of their segments are measured separately and the cutting entails some loss.

The centre of gravity of the right upper arm lay 13.4 cm below the centre of the head of the humerus and 17.2 cm above the axis of the humero-ulnar joint on the posterior border of the humerus. The centre of gravity of the right arm lay 28 cm below the centre of the head of the humerus and 2.6 cm above the axis of the humero-ulnar joint, 1.8 cm in front of the humerus, in the brachialis anticus muscle. The centre of gravity of the right forearm lay 10.9 cm below the axis of the humero-ulnar joint, 15.4 cm above the middle of the head of the capitate, and 1 cm in front of the interosseous membrane, in the flexor muscles, equidistant from the radius and ulna. The centre of gravity of the right forearm with the hand lay 17.8 cm below the axis of the humero-ulnar joint and 8.5 cm above the centre of the head of the capitate, 0.7 cm in front of the insertion of the interosseous membrane on the radius. The centre of gravity of the right hand, with flexed fingers lay 5.9 cm below the centre of the head of the capitate, about 1 cm above the head of the second metacarpal, equidistant, about 2 cm, from the second and third metacarpals, in the crease formed by the thumb and the palm of the hand, very close to the skin of the palm.

The centre of gravity of the left upper arm lay 13.7 cm below the centre of the head of the humerus and 16.5 cm above the axis of the humero-ulnar joint, immediately behind the humerus. The centre of gravity of the left arm lay 29.1 cm below the centre of the head of the humerus and 1.1 cm above the axis of the humero-ulnar joint, 1.4 cm in front of the middle of the humerus, in the brachialis anticus muscle. The centre of gravity of the left forearm lay 11 cm below the axis of the humero-ulnar axis and 16.1 cm above the centre of the head of the capitate, 1.1 cm in front of the interosseous membrane, equidistant from the radius and ulna, in the flexor muscles. The centre of gravity of the left forearm with the hand lay 17.6 cm below the axis of the humero-ulnar joint and 9.5 cm above the centre of the head of the capitate, 1 cm in front of the interosseous membrane, closer to the radius. The centre of gravity of the left hand, with the fingers somewhat less flexed than those of the right hand, lay 5.7 cm from the centre of the head of the capitate, 0.8 cm from the centre of the head of the third metacarpal, in the palm, 1.2 cm in front of the anterior border of the third metacarpal, equidistant from the second and fourth metacarpals. Here, the centre of gravity did not lie on the surface of the hand but in the tendon of the third flexor sublimis tendon.

The sizes and weights of the lower limbs of cadaver 3 are given in Table 4. To avoid misunderstandings, it must be stressed that the given lengths are not the lengths of the bones but the distances between the axes or centres of the joints. Therefore, the thigh appears to be too short and the lower leg too long.

The centre of gravity of the right thigh was 19.7 cm below the centre of the head of the femur and 22.3 cm above the axis of the knee, 2.2 cm behind the posterior border of the femur and somewhat medial. The centre of gravity of the right leg was 37.7 cm below the centre of the head of the femur, 4.3 cm above the axis of the knee, immediately at the posterior border of the femur. The centre of gravity of the right lower leg lay 18.7 cm below the axis of the knee and 24.3 cm above the axis of the ankle, 1 cm behind the interosseous membrane, a little closer to the tibia. The centre of gravity of the right lower leg with the foot lay 26.9 cm below the axis of the knee and 16.1 cm above the axis of the ankle, exactly at the insertion of the interosseous membrane onto the tibia. The centre of gravity of the right foot lay 11.4 cm from the posterior limit of the foot and 15.1 cm from its tip, in the angle formed by the inferior and lateral borders of the third cuneiform bone, close to its junction with the navicular bone.

The centre of gravity of the left thigh lay 19.5 cm below the centre of the head of the femur and 21.5 cm above the axis of the knee, 1.5 cm behind the linea aspera of the femur, somewhat medial. The centre of gravity of the whole left leg lay 38.5 cm below the centre of the head of the femur, 2.5 cm above the axis of the knee, 0.7 cm in front of the posterior border of the femur and somewhat lateral, in the bone itself. The centre of gravity of the lower left leg lay 17.7 cm below the axis of the knee, 25.2 cm above the axis of the ankle and 1 cm behind the interosseous membrane, closer to the tibia, in the flexor muscles. The centre of gravity of the left lower leg with the foot lay 26 cm below the axis of the knee, 19.9 cm above the axis of the ankle, at the insertion of the interosseous membrane onto the tibia. The centre of gravity of the left foot lay 11.8 cm from the posterior limit of the foot and 15.1 cm from its tip, in the angle between the third cuneiform and the cuboid bones, on the plantar side, close to the joint between the third cuneiform and the navicular bones.

Cadaver 4

This was the fresh, muscular cadaver of a manual worker who had hanged himself. The corpse was still in rigor mortis. The abdomen was a little sunken, so the border of the thorax, the xiphoid process, and the iliac crests appeared clearly. The cadaver was frozen lying supine. The forearms were slightly flexed and half pronated.

In order to obtain precise marks when sawing through the articular axes, rods were introduced and lines were marked on the corpse before freezing. The uniform hardness of the frozen body makes the osseous prominences more difficult to determine.

1. Rods were hammered dorsally through the centre of the humeral heads from the front.
2. The axis of the humero-ulnar joints was marked by a transverse line.
3. A needle was hammered through the middle of the head of the capitate.
4. The antero-superior iliac spines were marked.
5. The upper border of the two greater trochanters was marked by a transverse line on each side.
6. The axis of the knee joint was determined and marked by a transverse line on each side.
7. The axis of the ankle joint was determined and marked by a transverse line on each side.
   The projection of these lines and points onto the table made measurement of the body easier.

The cadaver was 168.8 cm long and weighed 56 090 g[11]. The centre of gravity of the whole body was found to lie in the lower pelvis, in the median plane of the

[11] Note, there is a discrepancy between the weight of the cadaver 4 as cited here and as given in Table 5. There was also a discrepancy between the text and Table 5 with regard to the weight of the right and left legs; in a section of the text that since has been deleted the weight of the former was originally given as 10 070 g, and that of the latter as 10 630 g. The discrepancies in the weights of the legs seem to result from the loss of material due to sawing, as mentioned by the authors. All these discrepancies occurred in the 1889 edition of this work.

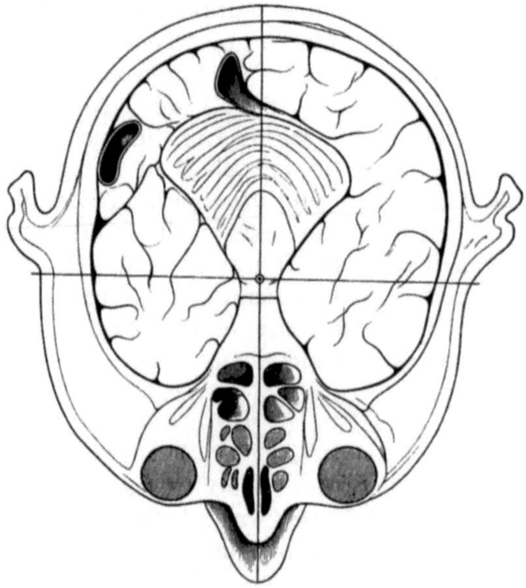

Fig. 1. Position of the centre of gravity in the head. ⊙ Centre of gravity. Scale: ⅓

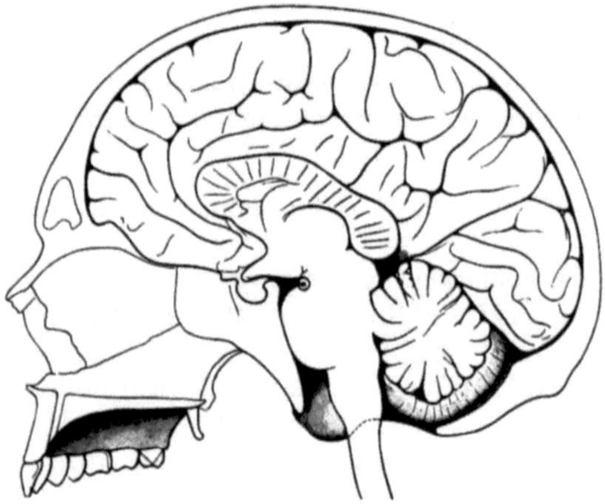

Fig. 2. Position of the centre of gravity in the head. ⊙ Centre of gravity. Scale: ⅓

body, 2.1 cm on the plumb line below the sacral promontory and 4.5 cm above the line joining the centres of the femoral heads, at the level of the upper border of the third sacral vertebra, 7 cm in front of the latter and 7 cm above the upper border of the symphysis.

The centre of gravity of the trunk without the head and arms lay at the upper border of the first lumbar vertebra, on its anterior aspect and 0.5 cm to the right of the median plane, 25.8 cm above the line connecting the centres of the femoral heads. The centre of gravity of the trunk with the head and arms lay almost in the median plane, 0.2 cm to the right of the latter, at the anterior border of the lower surface of the eleventh thoracic vertebra, 29 cm above the line joining the centres of the femoral heads.

The head was cut from below the chin backwards and upwards in order to separate the head from the neck as much as possible. Subsequent examinations showed that the three upper cervical vertebrae had remained with the skull. The centre of gravity of the head lay exactly in the median plane, 0.7 cm behind the dorsum sellae, exactly in the angle formed by the anterior border of the pons and the lamina perforata (Figs. 1, 2).

The measurements of the upper limbs are given in Table 5. In Table 5 the differences between the sums of the total weights, which result from the loss during sawing, have been compensated so that the weights can be directly used in subsequent calculation.

Since the lengths of the two upper limbs and their segments differed only by a few millimetres, we calculated the average values in order to obtain symmetrical conditions. However, we considered the weights separately. To give an accurate notion of the position of the centres of gravity in the upper and lower extremities, we have indicated these on cross sections of the cadaver (Figs. 3, 4).

Table 5. Data concerning cadaver 4

Part of body measured	Right		Left
Weight (g)			
Whole body		55 700	
Head		3 930	
Trunk without limbs		23 780	
Arm	3 520		3 710
Upper arm	1 730		2 020
Forearm with hand	1 790		1 690
Forearm without hand	1 300		1 240
Hand	490		450
Leg	10 110		10 650
Thigh	6 150		6 750
Lower leg with foot	3 960		3 900
Lower leg without foot	2 970		2 900
Foot	990		1 000
Length, height (cm)			
Length of whole cadaver		168.8	
Height of head [a]		21.3	
Distance between the chin and the centre of the line joining the centres of the humeral heads		10.1	
Distance between the line joining the centres of the humeral heads and the line joining the centres of the femoral heads		49	
Length of thigh	40		40
Length of lower leg	41.5		41.5
Height of foot	6.5		6.5
Length of foot	26.5		26.5
Distance between the centres of the femoral heads		17	
Length of upper arm	32		32
Length of forearm	27		27
Distance between the centres of the humeral heads [b]		36	

[a] From chin to vertex in vertical projection
[b] With the shoulders pushed forwards

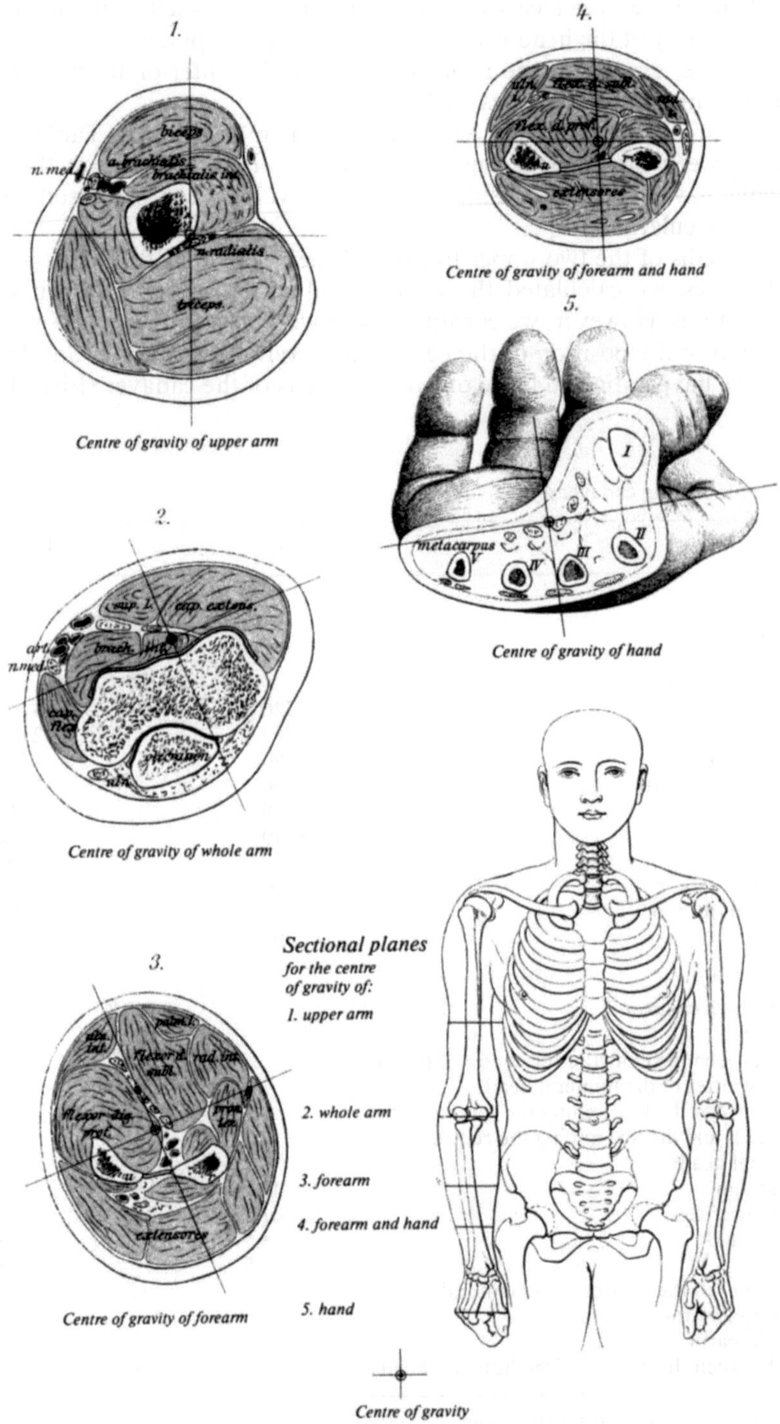

Fig. 3. Cross section with centres of gravity of the right arm of cadaver 4. Scale: 42/100

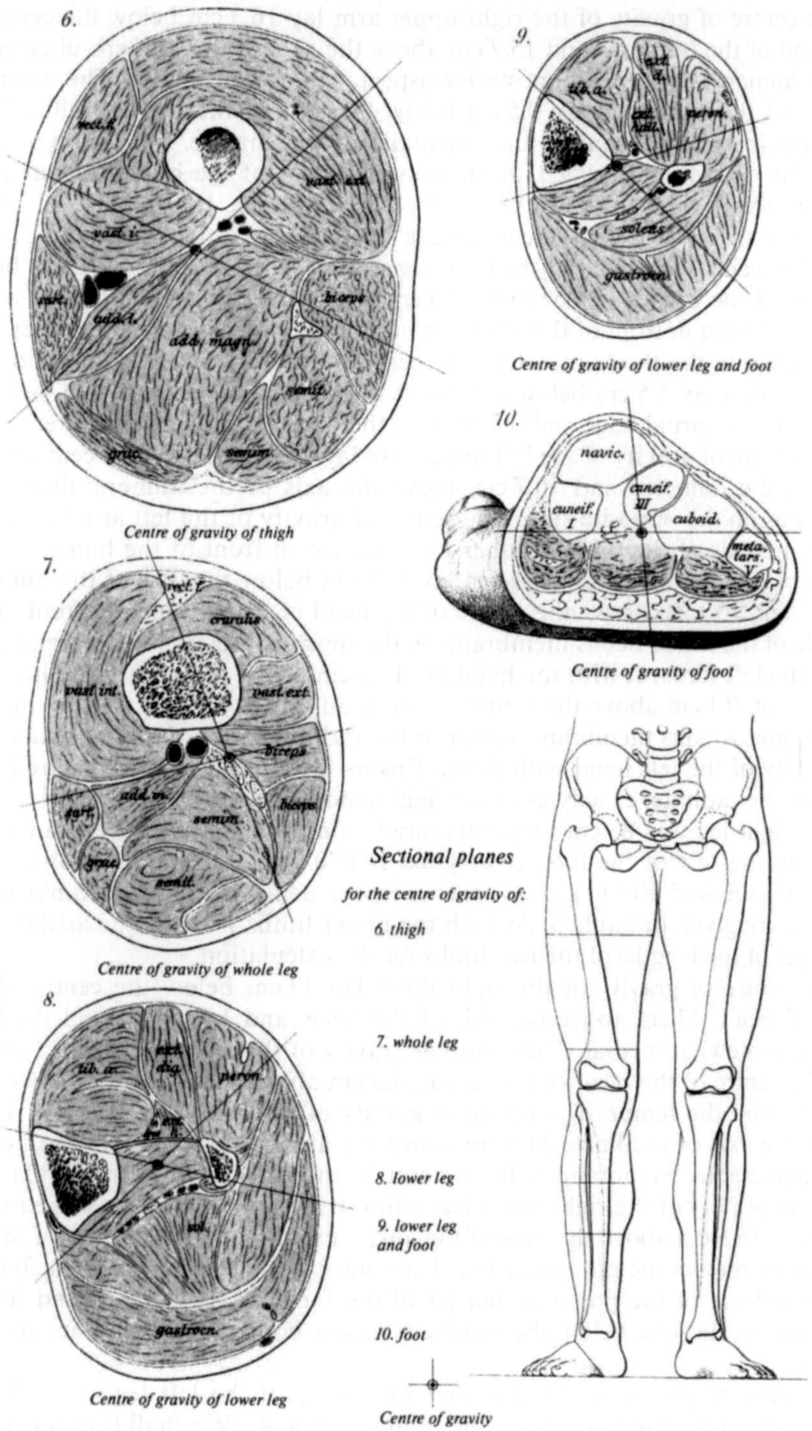

Fig. 4. Cross section with centres of gravity of the right leg of cadaver 4. Scale: 42/100

The centre of gravity of the right upper arm lay 16.3 cm below the centre of the head of the humerus and 15.7 cm above the axis of the humero-ulnar joint, in the humerus, close to its posterior aspect, somewhat medial. The centre of gravity of the right arm lay 0.5 cm below the axis of the humero-ulnar joint, 0.3 cm in front of the bone, in the brachialis anticus muscle. The centre of gravity of the right forearm lay 11.4 cm below the axis of the humero-ulnar joint, 15.6 cm above the centre of the head of the capitate, 1.5 cm in front of the middle of the interosseous membrane, a little closer to the radius than to the ulna. The centre of gravity of the right forearm with the hand lay 17.7 cm below the axis of the humero-ulnar joint, 9.3 cm above the centre of the head of the capitate, 0.5 cm in front of the interosseous membrane, closer to the radius than to the ulna, in the flexor muscles. The centre of gravity of the right hand with flexed fingers lay 5.5 cm below the centre of the head of the capitate, between the third metacarpal bone and the palm of the hand, 1 cm from the latter.

The centre of gravity of the left upper arm lay 15.3 cm below the centre of the head of the humerus and 16.7 cm above the axis of the humero-ulnar joint, exactly as on the opposite side. The centre of gravity of the left arm lay 0.5 cm above the axis of the humero-ulnar joint, 0.5 cm in front of the humerus. The centre of gravity of the left forearm lay 11.9 cm below the axis of the humero-ulnar joint, 15.1 cm above the centre of the head of the capitate, in front of the middle of the interosseous membrane, in the flexor muscles. The centre of gravity of the left forearm with the hand lay 17.9 cm below the axis of the humero-ulnar joint, 9.1 cm above the centre of the head of the capitate, 0.5 cm in front of the interosseous membrane, closer to the radius than to the ulna. The centre of gravity of the left hand with flexed fingers lay 5 cm below the centre of the head of the capitate, exactly as in the right hand.

The right leg was 88 cm long, measured from the centre of the head of the femur to the sole of the foot. It weighed 10 070 g. The left leg was also 88 cm long and weighed 10 630 g. The other measurements of the lower limbs in cadaver 4 are given in Table 5. As with the upper limbs we took the arithmetical averages of the lengths of the two limbs for the calculation.

The centre of gravity of the right thigh lay 17 cm below the centre of the femoral head, 23 cm above the axis of the knee, and 1.5 cm behind the linea aspera, somewhat medial. The centre of gravity of the right leg lay 35.5 cm below the centre of the head of the femur, 4.5 cm above the axis of the knee, and 1 cm behind the femur. The centre of gravity of the right lower leg lay 17 cm below the axis of the knee, 24.5 cm above the axis of the ankle, 0.7 cm behind the interosseous membrane, a little closer to the tibia than to the fibula. The centre of gravity of the right lower leg with the foot lay 25 cm below the axis of the knee, 16.5 cm above the axis of the ankle, exactly at the insertion of the interosseous membrane onto the tibia. The centre of gravity of the right foot lay 12 cm in front of the posterior border of the foot and 14.5 cm behind its tip, 3 cm above the sole, below the third cuneiform bone and close to its anterior border.

The relative positions of the centres of gravity of the left leg in width and depth were about the same as for the right leg (Fig. 4). We shall thus report only the heights of the centres of gravity in the left limb.

The centre of gravity of the left thigh lay 15 cm below the centre of the head of the femur and 25.4 cm above the axis of the knee. The centre of gravity of the left leg lay 33 cm below the centre of the head of the femur and 7 cm above the axis of the knee. The centre of gravity of the left lower leg lay 17.5 cm below the axis of the knee and 24 cm above that of the ankle. The centre of gravity of the left lower leg with the foot lay 25.5 cm below the axis of the knee and 16 cm above the axis of the ankle. The centre of gravity of the left foot was in exactly the same position as that of the right.

A remarkable relationship appeared between the centres of gravity of the different segments of the limbs and the axes of the joints. In the upper arm as well as in the thigh, the centre of gravity lay almost exactly on the straight line connecting the centre of the humeral head with the middle of the axis of the humero-ulnar joint or the centre of the femoral head with the middle of the axis of the knee. In the lower leg, the centre of gravity lay on or near the straight line that connects the middle of the axis of the knee and the centre of the ankle. In the forearm in mid-pronation, the centre of gravity lay on the straight line that passes through the centre of the head of the capitate and the middle of the axis of the humero-ulnar joint.

For the trunk, the curvature of the spine was measured as it was in the cadavers that had been frozen on their backs. The curvature as determined by Weber and Weber cannot be relied on because it does not correspond exactly to the normal conditions, having been determined on eviscerated cadavers. Removal of the viscera considerably alters the curvature of the spine, as demonstrated by Parow[12].

In the position that had been taken up during freezing, the centre of gravity of the trunk without the head and arms presented a particular relationship with the joints; it lay on or near a straight line connecting the centre of the occipito-atlantal joint and the middle of the line joining the centres of the two femoral heads.

According to our observations, the centre of gravity of all the segments of the limbs occurs on the line connecting the centres of the adjacent joints. This results in a considerable simplification of the calculations below. However, this rule of the position of the centres of gravity remains to be more precisely examined with regard to fluctuations between individuals. The data thus obtained enable us to determine an erect position of the body in which all the centres of gravity, except those of the feet, lie in one coronal plane. To do this, we have only to bring the centres of all the main joints into this plane. In order to establish whether such an erect position of the body is at all possible, we arranged a skeleton in this position and then compared it with a normal, well-built, living individual.

We took the dimensions of cadaver 4 and drew them to life-size on millimetre-paper in both lateral and antero-posterior view. We drew in the skeleton and outlines of the soft parts; we marked the centres of gravity found by direct measurement. We had marked the axes of the joints on the cadaver and we knew the position of the centres of gravity in the different segments in relation

[12] Parow, Virchow's Archiv, vol. 31

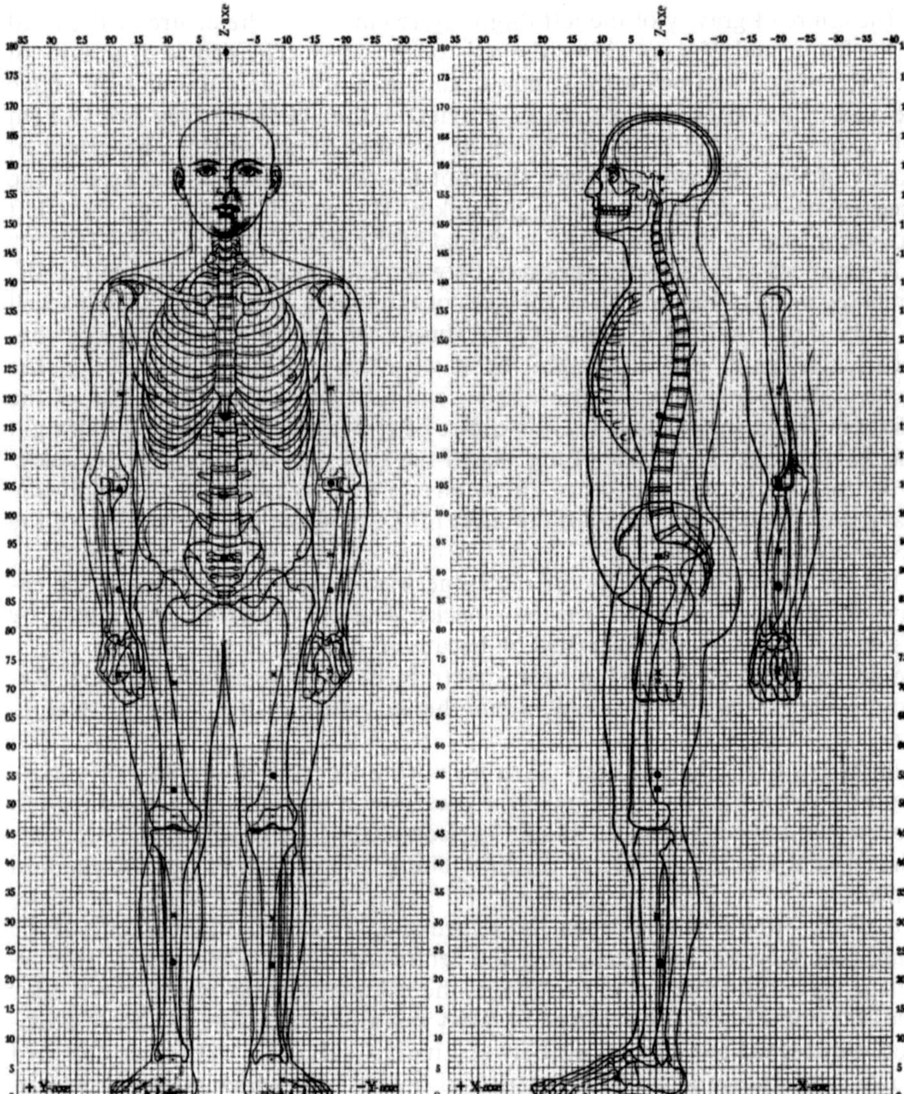

Fig. 5. Normal attitude of the human body with the centres of gravity as directly determined. × centre of gravity of segment; ⊙ centre of gravity of limb; ⊡ *S* centre of gravity of whole body

to the joints and bones. Therefore, we could accurately draw in the bones, their articular connections, and the centres of gravity (Fig. 5). The cross sections gave us the best opportunity to consider the proportions of the bones and soft parts in both width and length. We drew the body to life-size on millimetre-paper in order to enter the joints and centres of gravity into a rectangular system of co-ordinates. The two views give two projections on two planes perpendicular to each other and provide us with the three co-ordinates in the rectangular system of co-ordinates thus adopted.

The coronal plane is the plane *YZ*. The intersection of this coronal plane with the median plane of the body is the axis *Z* and its intersection with the horizontal ground on which the body stands erect is the axis *Y*. The perpendicular to the coronal plane at the intersection *O* of the two axes is the axis *X*. Axis *X* is positive forwards, axis *Y* positive to the right, and axis *Z* positive upwards from the origin *O* of the co-ordinates. The projection on the plane *XZ* gives the sagittal view and the horizontal ground surface coincides with the plane *XY*.

We assumed that the centre of gravity of the trunk lies exactly in the plane through the centres of the two hips and through the centre of the occipito-atlantal joint. We arranged the curvature of the spine accordingly. This corresponds to the evidence seen in median sections of the frozen cadavers. The position of the arms was somewhat modified. The arms were displaced forwards until the centres of the humeral heads and the centres of gravity of the different segments came to lie in the plane *YZ*. The displacement was so small that the position of the total centre of gravity could not have been altered significantly. The two views were reduced photographically (Fig. 5).

We now had to determine whether this position actually occurs in life and whether the horizontal position on the back could be transposed to an erect position. We cannot expect the curvature of the spine to remain exactly the same in both a horizontal and a vertical position. However, it is sufficient if the two positions are approximately similar. It is impossible to measure the curvature of the spine with complete accuracy in living individuals; even the co-ordinate measurer of Parow[13], which reproduces the curvature of the spine from the position of the spinal processes, gives only approximate results. However, we believe that we arrived at a satisfactory result in the following manner.

A well-built and strong young man, a soldier, was photographed in a lateral view. The centre of the joints had been marked on the skin as had the projection of the centre of gravity of the head on the lateral aspect of the latter. Two long plumb lines marked the plane *YZ*. The man was asked to modify his position until an observer standing at a distance could see all the marks of the articular points aligned in the plane of the plumb lines (plane *YZ*). The young man was photographed in this position, as shown in Fig. 6. The left elbow had to be flexed to avoid hiding the projection of the centre of the hip joint. This may have displaced the centre of gravity somewhat. However, we judged this displacement to be so small, compared with the advantage of seeing the axis of the hip, as to be negligible. Subsequent calculation showed this position to have displaced the centre of gravity only 7 mm upwards, 4 mm forwards, and 1 mm to the right.

A comparison of Figs. 5 and 6 shows good agreement between the actual attitude of the body and our construction, and it can be seen that the latter is not artificial but can be considered a natural attitude. In living individuals, the thorax protrudes a little more forwards and the lumbar lordosis is rather more pronounced than in our construction. But these discrepancies can be regarded as individual variations. The type of curvature of the spine and the position of the head are the same as in the construction in Fig. 5 even if the curvatures of

[13] Parow, Virchow's Archiv, vol. 31, p. 105

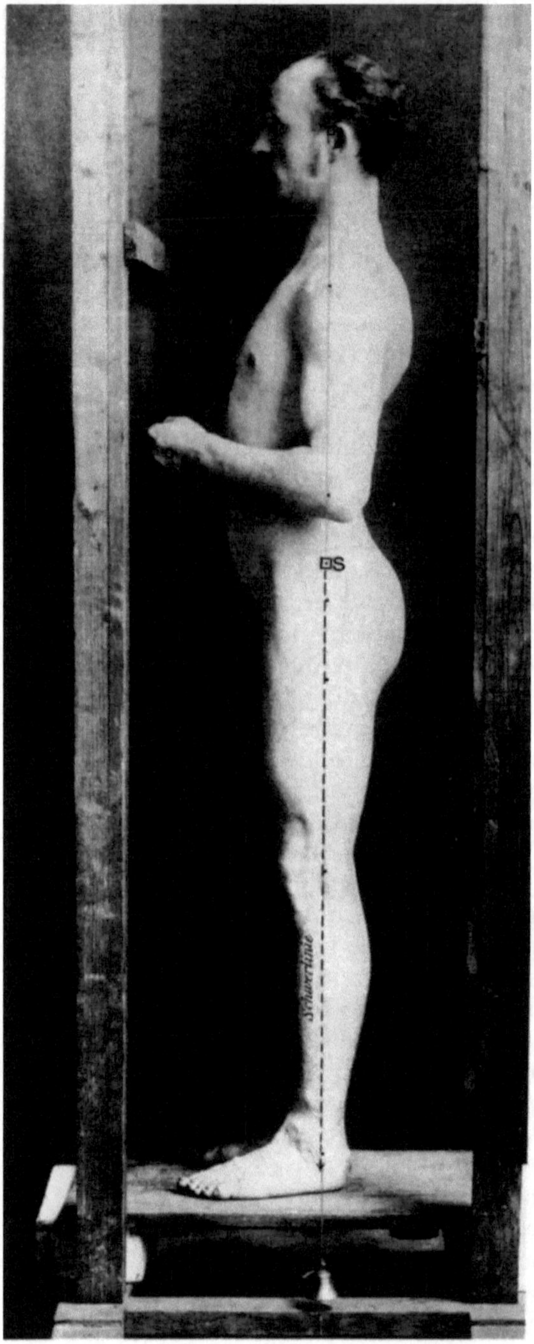

Fig. 6. Normal attitude, side view. · Projection of the centres of the joints; ⊡ projection of the centres of gravity of the head and hands; ⊡ *S* projection of the centre of gravity of the whole body. "Schwerlinie" = line of gravity

the different segments of the spine are somewhat more pronounced. However, the main point is that the attitude of the lower limbs and the position of the pelvis and head correspond exactly to our construction.

It is questionable whether this attitude is very comfortable. This problem was not considered in our studies; however, we shall come back to it later (pp. 64). Primarily we had one goal: to find a natural initial attitude that would be appropriate for measurements and calculations. We shall designate this attitude as the normal attitude.

So as to provide a control for our measurements, we calculated the positions of the centres of gravity of the whole body and limbs in the normal attitude from the positions of the centres of gravity of the different body segments. Entering the centres of the joints and the centres of gravity of the different body segments, in the normal attitude, into the system of co-ordinates gives the values in Tables 6 and 7 which can be read directly from Fig. 5.

The figures in parentheses in Table 7 in front of the words "head", "trunk", etc. relate to the numbering of the joints, tips of the feet, etc., as given in Table 6. They offer an easy way of briefly designating a well-determined body segment. For example, the thigh is designated as (3, 4) because it lies between the joints (3) and (4). This provides us with the possibility to distinguish the co-ordinates of the centres of the joints and of the centres of gravity with few words. For instance, the co-ordinates of the centre of the hip joint can be called x_3, y_3, z_3 and those of the thigh $x_{3,4}, y_{3,4}, z_{3,4}$. If we want to differentiate the right and left sides, we can designate the co-ordinates of the centre of the left hip as x'_3, y'_3, z'_3 and those of the left thigh as $x'_{3,4}, y'_{3,4}, z'_{3,4}$, whereas the prime is not used for the co-ordinates of the right side. We shall use these designations later. The co-ordinates of the centre of gravity of the whole body are expressed by x_0, y_0, z_0.

Table 6. Co-ordinates of the centres of the joints, vertex, tips of the feet, and lower ends of the hands with flexed fingers for cadaver 4

Co-ordinates		x	y	z
(1) Vertex		0	0	168.8
(2) Occipito-atlantal joint		0	0	154.0
(3) Hip joint	r	0	+ 8.5	88.0
	l	0	− 8.5	88.0
(4) Knee joint	r	0	+ 8.5	48.0
	l	0	− 8.5	48.0
(5) Ankle joint	r	0	+ 8.5	6.5
	l	0	− 8.5	6.5
(6) Tip of foot	r	+20.5	+13.0	0
	l	+20.5	−13.0	0
(7) Shoulder joint	r	0	+18.0	137.0
	l	0	−18.0	137.0
(8) Elbow joint	r	0	+18.0	105.0
	l	0	−18.0	105.0
(9) Wrist	r	0	+18.0	78.0
	l	0	−18.0	78.0
(10) Lower end of hand with flexed fingers	r	0	+18.0	67.5
	l	0	−18.0	67.5

Table 7. Co-ordinates of the centres of gravity for cadaver 4

Part of body[a]		x	y	z
(1, 2) Head		0	0	157.8
(2, 3) Trunk		0	+ 0.5	113.8
(3, 4) Thigh	r	0	+ 8.5	71.0
	l	0	− 8.5	72.5
(4, 5) Lower leg	r	0	+ 8.5	31.0
	l	0	− 8.5	30.5
(5, 6) Foot	r	+6.5	+10.3	3.0
	l	+6.5	−10.3	3.0
(7, 8) Upper arm	r	0	+18.0	120.7
	l	0	−18.0	121.7
(8, 9) Forearm	r	0	+18.0	93.6
	l	0	−18.0	93.1
(9, 10) Hand	r	0	+18.0	72.5
	l	0	−18.0	73.0
(1, 6, 7, 10) Whole body		0	0	92.5
(1, 3, 7, 10) Trunk+head+arms		0	0	117.0
(3, 6) Whole leg	r	0	+ 8.5	52.5
	l	0	− 8.5	55.0
(4, 6) Lower leg+foot	r	0	+ 8.5	23.0
	l	0	− 8.5	22.5
(7, 10) Whole arm	r	0	+18.0	104.5
	l	0	−18.0	105.5
(8, 10) Forearm+hand	r	0	+18.0	87.3
	l	0	−18.0	87.1

[a] See text for explanation of figures in parentheses

Table 8. Weights and adjusted weights of body segments of cadaver 4

Body segments		Weight in grams	Weight adjusted to whole-body weight of 10 000
(1, 2) Head		3 930	705.5
(2, 4) Trunk		23 780	4 270
(3, 4) Thigh	r	6 150	1 104
	l	6 750	1 212
(4, 5) Lower leg	r	2 970	533
	l	2 900	520.5
(5, 6) Foot	r	990	178
	l	1 000	179.5
(7, 8) Upper arm	r	1 730	310.5
	l	2 020	362.5
(8, 9) Forearm	r	1 300	233
	l	1 240	222.5
(9, 10) Hand	r	490	88
	l	450	81
(1, 6, 7, 10) Whole body		55 700	10 000

Since the weights of the body segments are necessary for the calculations, they are given again in Table 8. For these calculations, it is the relative proportions of the weights that are of importance. To calculate the total centre of gravity it is convenient to consider the total weight as 10 000. These adjusted weight figures for the various body segments are thus given beside the actual weights in Table 8.

We can designate the weight p of the right thigh as $p_{3,4}$, that of the left thigh as $p'_{3,4}$, etc.

Calculation of the Common Centre of Gravity of Parts of the Body and of the Whole Body from the Centres of Gravity and Weights of Their Different Segments

The forces of gravity are parallel. We shall thus explain how to combine two or more parallel forces into one resultant. In Fig. 7, A and B are the points of application of two parallel forces p_1 and p_2, represented by the segments AC and BD. Lengths and directions of these segments represent the magnitudes and directions of the two forces.

In the direction of the straight line connecting A and B, we add two equal forces of arbitrary magnitude but opposite direction. One is applied at point A

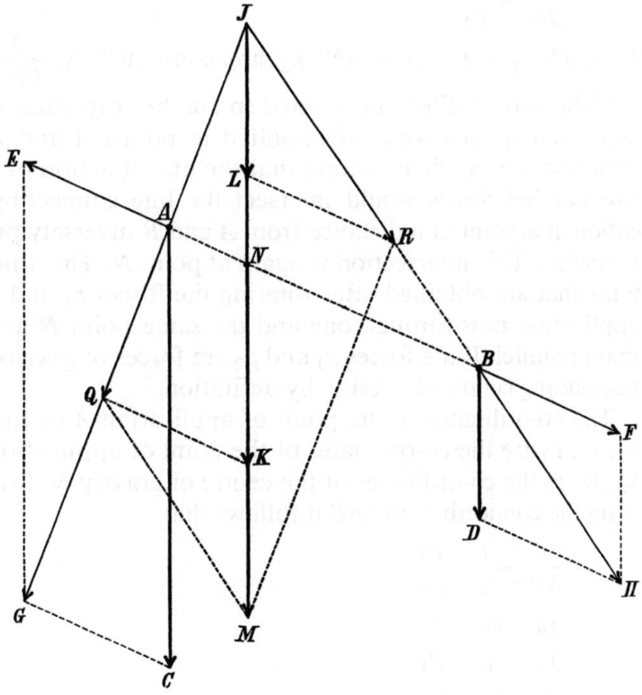

Fig. 7. Determination of the centre of gravity of a limb knowing the centres of gravity and weights of its two segments

and the other at B. Nothing is changed in the action of the initial system of forces. The two additional forces are represented by the segments AE and BF. Their magnitude is q. If the two forces acting at A are combined into one resultant AG, and the two forces acting at B into one resultant BH, these two resultants acting together are equivalent to the two initial forces p_1 and p_2. This remains true if the two resultants are displaced to the intersection J of their lines of action. J is the point of application of the two resultants. The two forces JQ and JR can be combined into one resultant JM, which is also the resultant of the two initial parallel forces.

The magnitude of this resultant is equal to the sum of the two parallel forces, p_1+p_2. It can be seen from Fig. 7 that $\Delta\, GCA \cong \Delta\, QKJ \cong \Delta\, RLM$ and that $\Delta\, HDB \cong \Delta\, RLJ \cong \Delta\, QKM$. Therefore, $JK = AC = p_1$ and $KM = BD = p_2$; thus $JM = p_1 + p_2$.

It results from the congruence of the triangles that JM is parallel to the two initial forces. Resultant JM intersects at N the straight line connecting the points of application A and B. The distances of this point N from A and B are inversely proportional to the parallel forces acting at A and B. As a consequence of the similarity of the triangles ANJ and QJK, it follows that:

$$\frac{AN}{JN} = \frac{q}{p_1}.$$

As a consequence of the similarity of the triangles BNJ and RLJ, it follows that:

$$\frac{BN}{JN} = \frac{q}{p_2}$$

Thus, $JN \cdot q = AN \cdot p_1 = BN \cdot p_2$, and consequently, $\dfrac{AN}{BN} = \dfrac{p_2}{p_1}$

If the two parallel forces acted in another direction but still parallel to each other and if they were still applied at points A and B, using the same construction, we can demonstrate that the line of action of the new resultant of the two parallel forces would intersect the line connecting their points of application at a point at a distance from A and B inversely proportional to the forces p_1 and p_2. This intersection is again at point N. The lines of action of the resultants that are obtained after rotating the forces p_1 and p_2 about their points of application pass through one and the same point N as long as these forces remain parallel. If the forces p_1 and p_2 are forces of gravity, the point N is the corresponding centre of gravity, by definition.

The co-ordinates of the point of application A of force p_1 are x_1, y_1, z_1 and x_2, y_2, z_2 are the co-ordinates of the point of application B of force p_2. We call x_0, y_0, z_0 the co-ordinates of the centre of gravity N. From the position of point N on the connecting line AB it follows that:

$$\frac{x_0 - x_1}{x_2 - x_0} = \frac{p_2}{p_1}$$

$$\frac{y_0 - y_1}{y_2 - y_0} = \frac{p_2}{p_1}$$

$$\frac{z_0 - z_1}{z_2 - z_0} = \frac{p_2}{p_1}$$

Therefore, the co-ordinates of the centre of gravity are:

$$x_0 = \frac{p_1 x_1 + p_2 x_2}{p_1 + p_2}$$

$$y_0 = \frac{p_1 y_1 + p_2 y_2}{p_1 + p_2}$$

$$z_0 = \frac{p_1 z_1 + p_2 z_2}{p_1 + p_2}$$

If there are three parallel forces p_1, p_2, p_3, the points of application of which have as co-ordinates x_1, y_1, z_1; x_2, y_2, z_2; and x_3, y_3, z_3, these three parallel forces can be combined into one resultant. This is arrived at by combining first forces p_1 and p_2 into one resultant and then combining this resultant with force p_3 into another resultant. We thus have finally to calculate the co-ordinates x_0', y_0', z_0' of the centre of gravity of the two forces $(p_1 + p_2)$ and p_3 the points of application of which have as co-ordinates x_0, y_0, z_0 and x_3, y_3, z_3. Thus:

$$x_0' = \frac{(p_1 + p_2)\, x_0 + p_3 x_3}{(p_1 + p_2) + p_3} = \frac{p_1 x_1 + p_2 x_2 + p_3 x_3}{p_1 + p_2 + p_3}$$

$$y_0' = \frac{(p_1 + p_2)\, y_0 + p_3 y_3}{(p_1 + p_2) + p_3} = \frac{p_1 y_1 + p_2 y_2 + p_3 y_3}{p_1 + p_2 + p_3}$$

$$z_0' = \frac{(p_1 + p_2)\, z_0 + p_3 z_3}{(p_1 + p_2) + p_3} = \frac{p_1 z_1 + p_2 z_2 + p_3 z_3}{p_1 + p_2 + p_3}.$$

From these three equations the rule appears according to which any number n of parallel forces $p_1, p_2, p_3, \ldots p_n$ can be combined into one resultant and the co-ordinates x_0, y_0, z_0 of the common centre of gravity of all the n gravity forces can be calculated.

$$x_0 = \frac{p_1 x_1 + p_2 x_2 + p_3 x_3 + \ldots p_n x_n}{p_1 + p_2 + p_3 + \ldots p_n} = \frac{\sum p_i x_i}{\sum p_i}$$

$$y_0 = \frac{p_1 y_1 + p_2 y_2 + p_3 y_3 + \ldots p_n y_n}{p_1 + p_2 + p_3 + \ldots p_n} = \frac{\sum p_i y_i}{\sum p_i}$$

$$z_0 = \frac{p_1 z_1 + p_2 z_2 + p_3 z_3 \ldots p_n z_n}{p_1 + p_2 + p_3 + \ldots p_n} = \frac{\sum p_i z_i}{\sum p_i}.$$

The values of x_0, y_0, z_0 are not altered if the actual weights p_i are replaced by figures proportional to these weights, for instance the figures obtained by assuming the total weight of the body to be 10 000. Then, in calculating the co-ordinates of the total centre of gravity of the body in any position, we are dealing with an easy divisor, 10 000, in the three equations, which is an appreciable advantage. Therefore, in the following calculations p_i will refer to the proportional figures as given on p. 30.

To calculate the co-ordinate z_0 of the centre of gravity of the whole body in the normal attitude using the co-ordinates z_i given on p. 30 and the proportional figures p_i of cadaver 4, we obtain the values of the products $p_i z_i$ as given

Table 9. Values of the products $p_i z_i$[a] for various parts of the body in cadaver 4

Head	111 327.9	Left foot	538.5
Trunk	485 926	Right upper arm	37 477.35
Right thigh	78 384	Left upper arm	44 116.25
Left thigh	87 870	Right forearm	21 808.8
Right lower leg	16 523	Left forearm	20 714.75
Left lower leg	15 875.25	Right hand	6 380
Right foot	534	Left hand	5 913
Total			933 388.8

[a] These products are given here because they will be cited in the equations in the text

in Table 9. Division by $\sum p_i = 10\,000$ gives the value of 93.3 cm for z_0 (rounded off to the first decimal place), whereas the value of the co-ordinate Z measured directly is 92.5 cm. Absolute accuracy obviously cannot be attained. Firstly, sawing the different parts of the body inevitably entails some weight loss. Secondly, the attitude of the body during measurement can not correspond exactly to the normal attitude. If we do not take into account the centres of gravity of the feet and if we thus calculate the mass that is supported by the feet and is balanced at the ankle joints, we obtain:

$$\sum p_i z_i = 933\,388.8 - (534 + 538.5) = 932\,316.3.$$

This must be divided by

$$10\,000 - (178 + 179.5) = 9642.5.$$

If we then designate these co-ordinates as x_0', y_0', z_0', the co-ordinate Z of the centre of gravity of the body without the feet is:

$$z_0' = \frac{932316.3}{9642.5} = 96.7 \text{ cm}.$$

To calculate the co-ordinate y_0 of the total centre of gravity of the body, we have the values of the products $p_i y_i$ as given in Table 10. Consequently $\sum p_i y_i = 29249.9 - 28563.1 = 686.8$ and $y_0 = \dfrac{686.8}{10000} = +0.07$ cm.

Table 10. Values of the products $p_i y_i$ for various parts of the body in cadaver 4

Part of the body	Right		Left
Head		0	
Trunk		+2 135	
Thigh	+ 9 384		− 10 302
Lower leg	+ 4 530.5		− 4 424.25
Foot	+ 1 833.4		− 1 848.85
Upper arm	+ 5 589		− 6 525
Forearm	+ 4 194		− 4 005
Hand	+ 1 584		− 1 458
Total	+29 249.9		− 28.563.1

Table 11. Co-ordinates of the centres of gravity of limbs in cadaver 4[a]

Body segment	$p_i z_i$		Adjusted weight		Co-ordinates	
	Right	Left	Right	Left	Right	Left
Whole leg						
Thigh	78 384	87 870	1 104	1 212		
Lower leg	16 523	15 875.25	533	520.5		
Foot	534	538.5	178	179.5	$z_{3,6} = \frac{95\,441}{1\,815} = 52.6$ cm	$z'_{3,6} = \frac{104\,283.75}{1\,912} = 54.5$ cm
Total	95 441	104 283.75	1 815	1 912	$y_{3,6} = +8.5$ cm	$y'_{3,6} = -8.5$ cm
					$x_{3,6} = 0$ cm	$x'_{3,6} = 0$ cm
Lower leg and foot						
Lower leg	16 523	15 875.25	533	520.5		
Foot	534	538.5	178	179.5	$z_{4,6} = \frac{17\,057}{711} = 24.0$ cm	$z'_{4,6} = \frac{16\,413.75}{700} = 23.5$ cm
Total	17 057	16 413.75	711	700	$y_{4,6} = +8.5$ cm	$y'_{4,6} = -8.5$ cm
					$x_{4,6} = +0$ cm	$x'_{4,6} = +0$ cm
Whole arm						
Upper arm	37 477.35	44 116.25	310.5	362.5		
Forearm	21 808.8	20 714.75	233	222.5		
Hand	6 380	5 913	88	81	$z_{7,10} = \frac{65\,666.15}{631.5} = 104.0$ cm	$z'_{7,10} = \frac{70\,744}{666} = 106.2$ cm
Total	65 666.15	70 744	631.5	666	$y_{7,10} = +18$ cm	$y'_{7,10} = -18$ cm
					$x_{7,10} = 0$ cm	$x'_{7,10} = 0$ cm
Forearm and hand						
Forearm	21 808.8	20 714.75	233	222.5		
Hand	6 380	5 913	88	81	$z_{8,10} = \frac{28\,188.8}{321} = 87.8$ cm	$z'_{8,10} = \frac{26\,627.75}{303.5} = 87.7$ cm
Total	28 188.8	26 627.75	321	303.5	$y_{8,10} = +18$ cm	$y'_{8,10} = -18$ cm
					$x_{8,10} = 0$ cm	$x'_{8,10} = 0$ cm

[a] These co-ordinate values and those in Table 12 should be compared with the values directly measured, given in Table 7

The calculation thus indicates a deviation of only 0.07 cm to the right of the median plane whereas direct measurement gave it in the median plane.

The co-ordinate Y of the centre of gravity of the body without the foot is:

$$y_0' = \frac{686.8 - (1833.4 - 1848.85)}{9642.5} = \frac{702.25}{9642.5} = +\,0.07 \text{ cm} .$$

These results are thus identical.

For the calculation of the co-ordinate x_0 of the total centre of gravity of the body in the normal attitude, only the products $p_i x_i$, which relate to the centres of gravity of the feet, are not equal to 0. Therefore:

$$x_0 = \frac{178 \cdot 6.5 + 179.5 \cdot 6.5}{10000} = \frac{2323.75}{10000} = 0.2 \text{ cm} .$$

This means that the total centre of gravity of the body in this attitude lies 2 mm in front of the coronal plane through all the partial centres of gravity except those of the feet. The co-ordinate X of the centre of gravity of the body without the feet is 0 since all x_i are equal to 0.

In addition to the co-ordinates of the total centre of gravity of the body, we have also calculated the co-ordinates of the centres of gravity of the limbs (Tables 11, 12); these centres of gravity had been determined experimentally (Table 7).

Table 12. Co-ordinates of the centres of gravity in cadaver 4

Body segment	$p_i z_i$	Adjusted weight	$p_i y_i$	Co-ordinates
Trunk and head				
Trunk	485 926	4 270		
Head	111 327.9	705.5		$z_{1,3} = \frac{597\,253.9}{4\,975.5} = 120$ cm
Total	597 253.9	4 975.5		$y_{1,3} = \frac{2\,135}{4\,975.5} = 0.4$ cm
Trunk, head, and arm				$x_{1,3} =$ 0 cm
Trunk	485 926	4 270		
Head	111 327.9	705.5		
Whole arm r	65 666.15	631.5		
Whole arm l	70 744	666		
Total	733 664.05	6 273		$z_{1,3,7,10} = \frac{733\,664.05}{6\,273} = 117.0$ cm
Trunk			2 135	
Right upper arm			5 589	
Right forearm			4 194	
Right hand			1 584	
Total			+13 502	
Head			0	
Left upper arm			− 6 525	
Left forearm			− 4 005	
Left hand			− 1 458	
Total			− 11 988	$y_{1,3,7,10} = \frac{13\,502 - 11\,988}{6\,273} = \frac{1514}{6273} = +0.2$ cm; $x_{1,3,7,10} = 0$ cm

Table 13. Average values of the co-ordinates of the centres of gravity in cadaver 4

Body segment	x	y[a]	z
Head	0	0	157.8
Trunk	0	0	113.8
Thighs	0	± 8.5	71.75
Lower legs	0	± 8.5	30.75
Feet	+6.5	±10.3	3
Upper arms	0	±18	121.2
Forearms	0	±18	93.35
Hands	0	±18	72.75

[a] The positive values of y are related to the right half of the body, the negative to the left

Table 14. Average weights and adjusted weights for cadaver 4

Body segment	Average weights (g)	Average adjusted weights	
		Average	Average to nearest figure
Whole body	55 700	10 000	10 000
Head	3 930	705.5	706
Trunk	23 780	4 270	4 270
Thigh	6 450	1 158	1 158
Lower leg	2 935	526.75	527
Foot	995	178.75	179
Upper arm	1 875	336.5	336
Forearm	1 270	227.75	228
Hand	470	84.5	84

Table 15. Co-ordinates of the centre of gravity for cadaver 4 assuming symmetrical body-mass distribution

Body segment	$p_i z_i$	Co-ordinates
Head	111 327.9	
Trunk	485 926	
Thighs	166 173	
Lower legs	32 395.125	
Feet	1 072.5	
Upper arms	81 567.6	
Forearms	42 520.925	
Hands	12 294.75	
Total	933 277.8	$\begin{cases} z_0 = \dfrac{933\,277.8}{10\,000} = 93.3 \text{ cm} \\ y_0 = = 0 \text{ cm} \\ x_0 = \dfrac{2 \times 6.5 \times 178.75}{10\,000} = 0.2 \text{ cm} \end{cases}$

Table 16. Co-ordinates of the centres of gravity for cadaver 4 assuming symmetrical mass distribution

Body segment	$p_i z_i$	Weight	Co-ordinate z	$p_i y_i$	Co-ordinate y	Co-ordinate x
Whole leg[a]						
Thigh	83 086.5	1 158		9 843		
Lower leg	16 197.5625	526.75		4 477.375		
Foot	536.25	178.75		1 841.125		
Total	99 820.3125	1 863.5	$z_{3,6}=\frac{99\,820.3125}{1\,863.5}=53.6$ cm	16 161.5	$y_{3,6}=\frac{16\,161.5}{1\,863.5}=8.7$ cm (for left leg, $y'_{3,6}=-8.7$ cm)	$x_{3,6}=0$ cm
Lower leg and foot[a]						
Lower leg	16 197.5625	526.75				
Foot	536.25	178.75				
Total	16 733.8125	705.5	$z_{4,6}=\frac{16\,733.8125}{705.5}=23.7$ cm		$y_{4,6}=\pm\ 8.5$ cm	$x_{4,6}=0$ cm
Whole arm[a]						
Upper arm	40 783.8	336.5				
Forearm	21 260.4625	227.75				
Hand	6 147.375	84.5				
Total	68 191.6375	648.75	$z_{7,10}=\frac{68\,191.6375}{648.75}=105.1$ cm		$y_{7,10}=\pm 18$ cm	$x_{7,10}=0$ cm
Forearm and hand[a]						
Forearm	21 260.4625	227.75				
Hand	6 147.375	84.5				
Total	27 407.8375	312.25	$z_{8,10}=\frac{27\,407.8375}{312.25}=87.8$ cm		$y_{8,10}=\pm 18$ cm	$x_{8,10}=0$ cm
Head, trunk, and both arms						
Head	111 327.9	705.5				
Trunk	485 926	4 270				
Upper arms	81 567.6	673				
Forearms	42 520.925	455.5				
Hands	12 294.75	169				
Total	733 637.175	6 273	$z_{1,3,7,10}=\frac{733\,637.175}{6\,273}=117.0$ cm		$y_{1,3,7,10}=0$ cm	$x_{1,3,7,10}=0$ cm
Head and trunk						
Head	111 327.9	705.5				
Trunk	485 926	4 270				
Total	597 253.9	4 975.5	$z_{1,3}=\frac{597\,253.9}{4\,975.5}=120.0$ cm		$y_{1,3}=0$ cm	$x_{1,3}=0$ cm

[a] Applies to both right and left limbs

If an equal and symmetrical distribution of the body mass is assumed, the arithmetical average of the co-ordinates of the centres of gravity in Table 7 and the weight figures in Table 8 must be taken. If we place the centre of gravity of the trunk exactly in the median plane, the average values of the co-ordinates of the centres of gravity are as in Table 13. The averages of the weights and those of the weights adjusted to a total weight of 10 000 are then as in Table 14.

If we assume an equal and symmetrical distribution of the body mass, the calculation of the co-ordinates of the centre of gravity gives the figures in Table 15.

Excluding the feet, $\sum p_i z_i$ becomes:

$$\begin{array}{rl} & 933\,277.8 \text{ (whole body)} \\ - & 1\,072.5 \text{ (feet)} \\ \hline = & 932\,205.3 \end{array}$$

The weight becomes:

$$\begin{array}{rl} & 10\,000 \text{ (whole body)} \\ - & 357.5 \text{ (feet)} \\ \hline = & 9642.5 \end{array}$$

Therefore: $z'_0 = \dfrac{932\,205.3}{9642.5} = 96.7 \text{ cm}; \quad y'_0 = 0 \text{ cm}; \quad z'_0 = 0 \text{ cm}.$

The co-ordinates of the centres of gravity of the different body segments have also been calculated assuming symmetrical mass distribution. They are given in Table 16.

The clear agreement between the calculated position of the centres of gravity (Tables 15, 16) and their positions as directly measured on the cadaver (Table 7) shows that our construction of the erect attitude cannot differ significantly from the position of the body segments in the supine position.

The positions of the centres of gravity are given in Tables 17–20 so as to provide an overall picture and to permit comparison with the results arrived at on the different cadavers. Table 17 gives the results in detail; Tables 18–20 give the relative distances of the centres of gravity from the centre of the joints and their averages. The length of each body segment is assumed to be 1 in these tables.

The relative distances of the centre of gravity of the trunk from the occipito-atlantal joint and from the line connecting the centres of the femoral heads were 0.609 and 0.391, respectively, if the distance between the occipito-atlantal joint and the line connecting the centres of the femoral heads is assumed to be 1.

If the relative distances of the limb segments are taken only to two decimal places, we obtain the data in Table 20.

These figures closely approximate the fractions four-ninths and five-ninths.

The centre of gravity of the different segments of the limbs thus lies at about the mid-point of these segments. Its distance from the centre or axis of the two adjacent joints is in the ratio of 4 : 5.

To determine the approximate position of the centre of gravity of any segment of a limb, one has to measure the distance between the two adjacent joint axes and divide this distance by 9. The centre of gravity lies four-ninths of the dis-

Table 17. Positions of the centres of gravity of the whole body and different body segments found in cadavers 2–4

	Cadaver 2	Cadaver 3	Cadaver 4
Total length (cm)	170	166	168.8
Total weight (g)	75 100	60 750	56 090[a]
Centre of gravity of whole body	Close to sacral promontory but below it	4 cm in front of lower border of 1st sacral vertebra, 0.2 cm to right	2.1 cm below sacral promontory 7 cm in front of upper border of 3[d] sacral vertebra
Centre of gravity of head	With neck. At the clivus, exactly in median plane at junction of sphenoid and occipital	Without neck. Behind dorsum sellae, exactly in median plane	Without neck. Exactly as in cadaver 3, 0.7 cm behind dorsum sellae, in median plane
Centre of gravity of trunk without arms and head	In anterior aspect of lower border of 1st lumbar vertebra (close to median plane)	In middle of 1st lumbar vertebra, 2 cm from anterior aspect, 1.4 cm from posterior aspect of body of vertebra, 0.3 cm to right of median plane	In superior aspect of 1st lumbar vertebra, at anterior border of vertebral body, 0.5 cm to right
Right leg	Centre of gravity in thigh, 39 cm from upper end, 5 cm from lower end, in middle of bone	Centre of gravity in thigh, 37.7 cm from upper end, 4.3 cm from lower end, in middle of bone	Centre of gravity in thigh, 35.5 cm from upper end, 4.5 cm from lower end, 1 cm behind bone
Left leg	38.5 cm from upper end of thigh, 4.8 cm from lower end, as right leg	38.5 cm from upper end of thigh, 2.5 cm from lower end, in bone itself, 0.7 cm in front of posterior aspect, somewhat lateral. Length: 50.7 cm	In thigh, 33 cm from upper end, 7 cm from lower end, as right leg
Right thigh	Length: 44 cm; weight: 7650 g. Centre of gravity 19 cm from upper end, 25 cm from lower end, 1.5 cm behind linea aspera, somewhat medial	Length: 42 cm; weight: 6690 g. Centre of gravity 19.7 cm from upper end, 22.3 cm from lower end, 2.2 cm behind linea aspera	Length: 40 cm (average); weight: 6150 g. Centre of gravity 17 cm from upper end, 23 cm from lower end, 1.5 cm behind linea aspera
Left thigh	Length: 43.3 cm; weight: 7300 g. Centre of gravity 19.3 cm from upper end, 24 cm from lower end	Length: 41 cm; weight: 6220 g. Centre of gravity 19.5 cm from upper end, 21.5 cm from lower end, 1.5 cm behind linea aspera	Length: 40 cm (average); weight: 6750 g. Centre of gravity 15.5 cm from upper end, 24.5 cm from lower end, as right thigh

Right lower leg with foot	Centre of gravity in lower leg 24.6 cm from upper end, 16.8 cm from lower end, immediately behind insertion of interosseous membrane onto tibia. Length: 49.2 cm, lower leg + height of foot	Centre of gravity in lower leg, 26.9 cm from upper end, 16.1 cm from lower end, exactly at insertion of interosseous membrane onto tibia. Length: 50.7 cm, lower leg + height of foot	Centre of gravity in lower leg, 25 cm from upper end, 16.5 cm from lower end, exactly at insertion of interosseous membrane onto tibia. Length: 48 cm, lower leg + height of foot
Left lower leg with foot	25.5 cm from upper end, 16.3 cm from lower end, otherwise as right leg. Length: 49.3 cm	26 cm from upper end, 19.9 cm from lower end, otherwise as right leg. Length: 50.6 cm	25.5 cm from upper end, 16 cm from lower end, otherwise as right leg. Length: 48 cm
Right foot	Length: 28.5 cm; height: 7.8 cm; weight: 1100 g. Centre of gravity 11.5 cm from posterior limit, 17 cm from anterior aspect, 6.5 cm in front of axis of ankle joint, at anterior border of navicular, between the 2nd and 3rd cuneiforms	Length: 26.5 cm; height: 7.7 cm; weight: 1060 g. Centre of gravity 11.4 cm from posterior limit, 15.1 cm from anterior limit in angle formed by lower and lateral borders of 3rd cuneiform, close to navicular	Length: 26.5 cm; height: 6.5 cm; weight: 990 g. Centre of gravity 12 cm from posterior limit, 14.5 cm from anterior limit, below 3rd cuneiform, close to anterior border, 3 cm above sole
Left foot	Length: 28.3 cm; height: 7.5 cm; weight: 1160 g. Centre of gravity 12 cm from posterior limit, 16.3 cm from anterior limit, otherwise as right foot	Length: 26.9 cm; height: 7.7 cm; weight: 1090 g. Centre of gravity 11.8 cm from posterior limit, 15.1 cm from anterior limit, on anterior surface of navicular between 3rd cuneiform and cuboid	Length: 26.5 cm; height: 6.5 cm; weight: 1000 g. Centre of gravity 12 cm from posterior limit, 14.5 cm from anterior limit, otherwise as right foot
Right lower leg	Length: 41.4 cm; weight: 3210 g. Centre of gravity 17.4 cm from upper end, 24 cm from lower end, 1 cm behind interosseous membrane	Length: 43 cm; weight: 2870 g. Centre of gravity 18.7 cm from upper end, 24.3 cm from lower end, a little behind interosseous membrane	Length: 41.5 cm (average); Weight: 2970 g. Centre of gravity 17 cm from upper end, 24.5 cm from lower end behind interosseous membrane
Left lower leg	Length: 41.8 cm; weight: 3320 g. Centre of gravity 17.4 cm from upper end, 24.4 cm from lower end, as right leg	Length: 42.9 cm; weight: 2880 g. Centre of gravity 17.7 cm from upper end, 25.2 cm from lower end, 1 cm behind interosseous membrane	Length: 41.5 cm (average); weight: 2900 g. Centre of gravity 17.5 cm from upper end, 24 cm from lower end, as right leg

Table 17. (continued)

	Cadaver 2	Cadaver 3	Cadaver 4
Right arm	Centre of gravity in arm, somewhat below axis of humero-ulnar joint, 1.5 cm in front of bone	Centre of gravity in upper arm, 28 cm from upper end, 2.6 cm from lower end, 1.8 cm in front of humerus	Centre of gravity in upper arm, 0.5 cm below axis of humero-ulnar joint, 0.3 cm in front of bone
Left arm	As right arm	Centre of gravity 29.1 cm from upper end of upper arm, 1.1 cm from lower end, 1.4 cm in front of middle of humerus	Centre of gravity 0.5 cm above axis of humero-ulnar joint, 0.5 cm in front of bone
Right upper arm	Length (to joint space of humero-ulnar joint): 31.7 cm; weight: 2850 g. Centre of gravity 14.5 cm from upper end (centre of humeral head), 17.2 cm above axis of humero-ulnar joint, in medullary cavity of humerus, at posterior border	Length: 30.6 cm; weight: 1990 g. Centre of gravity 13.4 cm from upper end, 17.2 cm from lower end, at posterior surface of humerus	Length: 32.0 cm; weight: 1730 g. Centre of gravity 16.3 cm from upper end, 15.7 cm from lower end, in humerus close to posterior border
Left upper arm	Length: 31.5 cm (to joint space of humero-ulnar joint): weight: 2560 g. Centre of gravity 13.3 cm from upper end, 18.2 cm from lower end, in medullary cavity	Length: 30.2 cm; weight: 1880 g. Centre of gravity 13.7 cm from upper end, 16.5 cm from lower end, on posterior surface of humerus	Length: 32 cm; weight: 2020 g. Centre of gravity 15.3 cm from upper end, 16.7 cm from lower end, as right arm
Right forearm with hand	Centre of gravity in forearm, 19 cm from axis of humero-ulnar joint, 10.5 cm from head of capitate, 0.5 cm in front of insertion of interosseous membrane onto radius. Length from axis of humero-ulnar axis to lower end of flexed fingers: 41.5 cm	Centre of gravity in forearm, 17.8 cm from upper end, 8.5 cm from lower end, 0.7 cm from insertion of interosseous membrane onto radius. Length from axis of humero-ulnar joint to lower end of flexed fingers: 37.5 cm	Centre of gravity in forearm, 17.7 cm from upper end, 9.3 cm from lower end in front of interosseous membrane, closer to radius. Length from axis of humero-ulnar joint to lower end of flexed fingers: 37.5 cm
Left forearm with hand	Centre of gravity 19 cm from upper end, 10.5 cm from lower end of forearm, 1 cm in front of insertion of interosseous membrane, onto radius. Length: 41.5 cm	Centre of gravity 17.6 cm from upper end, 9.5 cm from lower end of forearm, 1 cm in front of interosseous membrane, closer to radius. Length: 38.0 cm	Centre of gravity 17.9 cm from upper end, 9.1 cm from lower end of forearm, 0.5 cm in front of interosseous membrane, closer to radius. Length: 37.5 cm

Right forearm	Length: 29.5 cm; weight: 1700 g. Centre of gravity 12.5 cm from upper end, 17 cm from lower end, 1 cm in front of interosseous membrane	Length: 26.3 cm; weight: 1050 g. Centre of gravity 10.9 cm from upper end, 15.4 cm from lower end, 1 cm in front of interosseus membrane	Length: 27 cm (average); weight: 1300 g. Centre of gravity 11.4 cm from upper end, 15.6 cm from lower end, 1.5 cm in front of middle of interosseous membrane
Left forearm	Length: 29.5 cm; weight: 1600 g. Centre of gravity 12.4 cm from upper end, 17.1 cm from lower end, 1 cm in front of interosseous membrane	Length: 27.1 cm; weight: 1120 g. Centre of gravity 11.0 cm from upper end, 16.1 cm from lower end, 1.1 cm in front of interosseous membrane	Length: 27 cm (average); weight: 1240 g. Centre of gravity 11.9 cm from upper end, 15.1 cm from lower end, as right forearm
Right hand with fingers slightly flexed	Weight: 670 g. Centre of gravity 5.5 cm from upper end (centre of head of capitate), in palm, 1 cm in front of centre of head of 3rd metacarpal	Weight: 500 g. Centre of gravity 5.9 cm from upper end, 1 cm in front of head of 2nd metacarpal, close to skin of palm	Weight: 490 g. Centre of gravity 5.5 cm from uppper end, between 3rd metacarpal and palm, 1 cm from latter
Left hand with fingers slightly flexed	Weight: 620 g. Centre of gravity as right hand	Weight: 470 g. Centre of gravity 5.7 cm from upper end, 0.8 cm in front of centre of head of 3rd metacarpal	Weight: 450 g. Centre of gravity 5 cm from upper end. Otherwise, as right hand

[a] See p. 21

Table 18. Relative distances of the centres of gravity from the axes or centres of the joints[a]

Body segment		Cadaver 2[b]		Cadaver 3		Cadaver 4	
		From upper end	From lower end	From upper end	From lower end	From upper end	From lower end
Upper arm	r	–	–	0.438	0.562	0.509	0.491
	l	–	–	0.454	0.546	0.478	0.522
Forearm	r	–	–	0.414	0.586	0.422	0.578
	l	–	–	0.406	0.594	0.441	0.559
Thigh	r	0.432	0.568	0.469	0.531	0.425	0.575
	l	0.446	0.554	0.476	0.524	0.3875	0.6125
Lower leg	r	0.420	0.580	0.435	0.565	0.410	0.590
	l	0.416	0.584	0.413	0.587	0.422	0.578
Foot[c]	r	0.404	0.596	0.430	0.570	0.453	0.547
	l	0.424	0.576	0.439	0.561	0.453	0.547
Forearm and hand[d]	r	–	–	0.475	0.525	0.472	0.528
	l	–	–	0.463	0.537	0.477	0.523
Lower leg and foot[e]	r	0.500	0.500	0.531	0.469	0.521	0.479
	l	0.517	0.483	0.514	0.486	0.531	0.469

[a] The distance between the axes or centres of the joints adjacent to each body segment is assumed to be 1

[b] The relative distances are not given for the upper limbs of cadaver 2 because, in this cadaver, the saw cut passed through the humero-ulnar joint and not through the axis. The figures thus would allow no comparison with those of cadavers 3 and 4

[c] The antero-posterior length = 1

[d] The distance between the axis of the humero-ulnar joint and the lower end of the flexed fingers = 1

[e] The distance between the axis of the knee and the sole = 1

tance between the two axes, below the axis of the proximal joint, on the straight line connecting the centres of the two adjacent joints. In the thigh and upper arms, the centre of gravity lies close to the posterior border of the bone. In the forearm and lower leg it lies on average on the flexor aspect of the segment, 1 cm from the middle of the interosseous membrane.

The direct measurements of the distances of the centres of gravity of the different segments of limbs from the joints, as carried out by Harless, are in satisfactory agreement with our own results as long as the averages of the different observations are considered. Complete agreement obviously cannot be expected because individual variations occur and the softness of his material constituted a source of error. In our own cadavers, individual differences appeared. However, they were smaller than the differences in the two series of measurements of Harless. This demonstrates that the differences in the Harless measurements cannot be attributed only to individual variations, as previously suspected.

[14] Harless, op. cit. pp. 93, 271

Table 19. Arithmetical averages of the relative distances of the centre of gravity of each body segment from the axes or centres of the adjacent joints in cadavers 2–4

Body segment		From upper end		From lower end	
		Average of relative distances of one side	Average of relative distances of both sides	Average of relative distances of one side	Average of relative distances of both sides
Upper arm	r	0.4735	0.470	0.5265	0.530
	l	0.446		0.534	
Forearm	r	0.418	0.421	0.582	0.579
	l	0.4235		0.5765	
Thigh	r	0.442	0.439	0.558	0.561
	l	0.4365		0.5635	
Lower leg	r	0.422	0.4195	0.578	0.5805
	l	0.417		0.583	
Foot	r	0.429	0.434	0.571	0.566
	l	0.439		0.561	
Forearm and hand	r	0.4735	0.472	0.5265	0.528
	l	0.470		0.530	
Lower leg and foot	r	0.517	0.519	0.483	0.481
	l	0.521		0.479	

Table 20. Averages from Table 19 taken to two decimal places

Body segments	From upper end	From lower end
Upper arm	0.47	0.53
Forearm	0.42	0.58
Thigh	0.44	0.56
Lower leg	0.42	0.58
Foot	0.43	0.57

Table 21. Relative distances of the centres of gravity from the joints in the cadavers "Graf" and "Kefer" of Harless

	"Graf"		"Kefer"		Average of both corpses	
	From upper end	From lower end	From upper end	From lower end	From upper end	From lower end
Upper arm	0.485	0.515	0.429	0.571	0.457	0.543
Forearm	0.439	0.561	0.410	0.590	0.4245	0.5755
Thigh	0.468	0.532	0.424	0.576	0.446	0.554
Lower leg	0.360	0.640	0.469	0.531	0.4145	0.5855
Foot	0.460	0.540	0.436	0.564	0.448	0.552

We have taken the results of Harless[14], reduced the length of the limb segments to 1, and calculated the average of the two sides to allow for a comparison with our own results (Table 21). The figures in Table 21 should be compared with our own data in Table 19. The relative distances in the results of Harless are also on average very close to the fractions four-ninths and five-ninths.

The position of the total centre of gravity of the body appears as the main difference between the results of Meyer, obtained by direct measurement, and ours. According to Meyer, in the erect position the total centre of gravity lies some distance behind the hip joints, in the second sacral vertebra or in the sacral canal. We never obtained this result in our own measurements. However, it must be stressed that we determined the centre of gravity with the cadaver lying in the supine position and subsequently constructed an erect position, which appeared to be possible when tested with living individuals, whereas Meyer and Horner carried out their observations directly on living subjects. It is questionable whether such an attitude can be imposed on the body that its centre of gravity comes to lie behind the hips, in the sacral canal, or even further posteriorly. The body would have to be arched with the shoulders considerably retracted so that it would form an arch ventrally convex. The centre of gravity would then come to lie behind the hip joints; the further from the hip joints the centre of gravity would come to lie, the more the body would be arched. In any case, the attitude described by Meyer cannot be used as the initial attitude for measurement since it is certainly not normal; neither is it "military" as claimed by Meyer. It still remains to be demonstrated that this attitude is at all stable. We shall return to this point later.

Meyer also claims[15] that the angle formed anteriorly by the lower leg and the foot, increased by lateral rotation of the foot, hinders the flexion forwards of the lower leg on the foot. This is helped by simultaneous hindrance in the knee and in the hip. This may be true for the ankle joint but not for the whole foot. If the thighs are firmly fixed, the spine can be arched quite as far backwards, as illustrated by Meyer. The trunk is then balanced by the ilio-femoral ligaments of Bertin. However, in the erect position, even with considerable external rotation of the feet, the movement of the leg is not hindered by the ankle joint. Sufficient mobility persists in the tarsus to allow for a considerable range of movement, even if the hips and knees are locked. Anyone can observe this on his own body. If there were no such mobility in the tarsus, it would be quite impossible to walk on a steep surface.

[15] Meyer (1853) Müllers Archiv, p. 25

Determining the Position of the Centre of Gravity in the Living Body in Different Attitudes and with Different Loads

Our measurements and deductions also enable us to determine the position of the centre of gravity in the living body. This requires entering the body into a tridimensional system of co-ordinates. Two projections of the body are sufficient, most conveniently on a sagittal and a coronal plane. These projections are reliably obtained by photographing a side and a front view. The plane XY of the rectangular system of co-ordinates is the horizontal ground. Axis Z is the vertical line through the middle of the line connecting the centres of the two hip joints. Axis X is directed postero-anteriorly, and axis Y transversely. The rectangular system of co-ordinates is thus complete if we add that axis X is positive forwards, axis Y is positive to the right and axis Z is positive upwards.

To orientate the body precisely for the photographs, we used two plumb lines, which gave the direction of axis Z in all the pictures. We used two plumb lines simultaneously to determine the plane YZ in the normal attitude. The centres of the joints were marked on the photographs thus obtained according to their projection onto the skin, which was arrived at by measuring the living subject in the corresponding attitude. Marking these points on the skin before taking the photographs led to inaçcuracies when the attitude was altered because of the mobility of the skin. Therefore, this method could only be used for the normal attitude. For all the other attitudes subsequently taken up by the body, the projections of the centres of the joints onto the skin had to be determined again.

The centres of the joints were also measured in the case of the elbows and knees since it appeared that the centre of gravity of a body segment always lies approximately on the line connecting the centres of the adjacent joints. Since we knew the distance of the centre of gravity of any body segment from the centres of the adjacent joints (Table 18), from the co-ordinates of these articular centres we could calculate the co-ordinates of the centre of gravity lying in between.

If x_i, y_i, z_i and x_k, y_k, z_k are the co-ordinates of the centres of the two joints adjacent to a body segment and if ε_i and ε_k are the relative distances of the centre of gravity from the centres of the adjacent joints, the following equations can be used to calculate the co-ordinates $x_{i,k}$, $y_{i,k}$, $z_{i,k}$ of the centre of gravity of the body segment:

$$\frac{x_i - x_{i,k}}{x_{i,k} - x_k} = \frac{\varepsilon_i}{\varepsilon_k}$$

$$\frac{y_i - y_{i,k}}{y_{i,k} - y_k} = \frac{\varepsilon_i}{\varepsilon_k}$$

$$\frac{z_i - z_{i,k}}{z_{i,k} - z_k} = \frac{\varepsilon_i}{\varepsilon_k}$$

and consequently:

$$(\varepsilon_i + \varepsilon_k)\, x_{i,k} = \varepsilon_k\, x_i + \varepsilon_i\, x_k$$

$$(\varepsilon_i + \varepsilon_k)\, y_{i,k} = \varepsilon_k\, y_i + \varepsilon_i\, y_k$$

$$(\varepsilon_i + \varepsilon_k)\, z_{i,k} = \varepsilon_k\, z_i + \varepsilon_i\, z_k\,.$$

The distances ε_i and ε_k are relative to the length 1 of the body segment, thus: $\varepsilon_i + \varepsilon_k = 1$, and it follows that:

$$x_{i,k} = \varepsilon_k\, x_i + \varepsilon_i\, x_k$$

$$y_{i,k} = \varepsilon_k\, y_i + \varepsilon_i\, y_k$$

$$z_{i,k} = \varepsilon_k\, z_i + \varepsilon_i\, z_k\,.$$

As relative distances ε_i, ε_k, we use the arithmetical averages of their values in the three determinations of the centre of gravity, brought to two decimal places as in Table 20. We thus obtaine the data in Table 22. ε_i relates to the proximal joint. The relative distances for the trunk have not been given in Table 22 because the trunk does not constitute an unalterable mass as do the different segments of the limbs. However, the variations of its shape in the attitudes of the body that we analysed are not considerable. Therefore, the relative distances arrived at for the normal attitude can be used without significant error, as was demonstrated by subsequent calculation. The occipito-atlantal joint was 66 cm from the middle of the line connecting the centres of the two hip joints; the distance of the centre of gravity from the occipito-atlantal joint was found to be 40.2 cm and it was 25.8 cm from the middle of the line connecting the centres of the two hip joints. This enables us to calculate ε_i and ε_k given in Table 22. We also assumed that the centre of gravity of the trunk always lies exactly in the median plane in the attitudes analysed.

The values ε_i and ε_k are not indicated for the hand because of the considerable variations in the attitude of the hand and fingers. However, the co-ordinates of the centre of gravity of the hand can be recognized fairly reliably in the pictures and directly measured.

To determine the weights of the different body segments in the living individuals, it would not be accurate to take the averages of the two determinations of Harless and our own results since Harless worked on bloodless corpses. Similarly, averages of our own three studies would not provide greater accuracy because the number of cadavers is much too small. It would seem more appropriate to take the weights of the cadaver that in size and weight most closely resembles the living subject under observation. The size and weight of cadaver 4 corresponded fairly well to those of our living subject. The difference in weight, however, was 3 kg. It was thus necessary to adjust the 55 700 g total weight of cadaver 4 to 58 700 g, the weight of our living subject. To do so, the weights of cadaver 4 had to be multiplied by a factor of: $\dfrac{58\,700}{55\,700} = 1.05386$.

The weights of the body segments of our living subject on which the calculations will be based are given Table 23 (to the nearest 10 g).

Table 22. Distances of the centre of gravity from the two adjacent joints

Body segment	ε_i	ε_k
Trunk	0.61	0.39
Thigh	0.44	0.56
Lower leg	0.42	0.58
Foot	0.43	0.57
Upper arm	0.47	0.53
Forearm	0.42	0.58

Table 23. Calculated weights of body segments of living subject

Body segment	Weight (g)
Total weight	58 700
Head	4 140
Trunk	25 060
Thigh	6 800
Lower leg	3 090
Foot	1 050
Upper arm	1 980
Forearm	1 340
Hand	490

The following attitudes of the same subject were photographed.

- Fig. 6 shows the normal attitude, as described on p. 27;
- Fig. 8 depicts the comfortable attitude, adopted spontaneously, without any external influence;
- Fig. 9 represents the military attitude.

A number of attitudes were then studied with the subject bearing military equipment. Figures 10–15 show attitudes adopted by the naked subject without a knapsack:

- Figs. 10 and 11 show the subject presenting arms;
- Figs. 12 and 13 demonstrate the subject in the shooting attitude;
- Figs. 14 and 15 depict the subject holding the rifle at arm's length.

Finally, pictures were taken representing the attitudes of the naked subject with full regulation equipment:

- Figs. 16 and 17 show the subject standing to attention at "shoulder arms";
- Figs. 18 and 19 show the subject in the shooting attitude.

For attitudes 1–3, the side views were sufficient because of the symmetry of the body. For the other attitudes, the front view was necessary in addition to the side view. In attitudes 1 and 2, the left forearm had to be flexed somewhat at the elbow so as to make the mark on the hip visible; in the military attitude, a slight displacement of the arm forward was sufficient. This required a slight correction to be made in the calculation so as to render the attitude symmetrical again.

The co-ordinates of the centres of the joints were determined from the photographs as follows. The centres of the joints were determined on the subject in the attitude under study, projected onto the skin, and marked on the photograph. Transparent millimetre paper was then laid on the photograph and orientated in such a way that the axis Z, corresponding to the plumb lines passed through the middle of the line connecting the centres of the two hip joints and the origin of the co-ordinates was on the horizontal ground. In this way, in attitudes 4–8, the three co-ordinates could be found from the two views, taking into account the shortenings and lengthenings due to perspective.

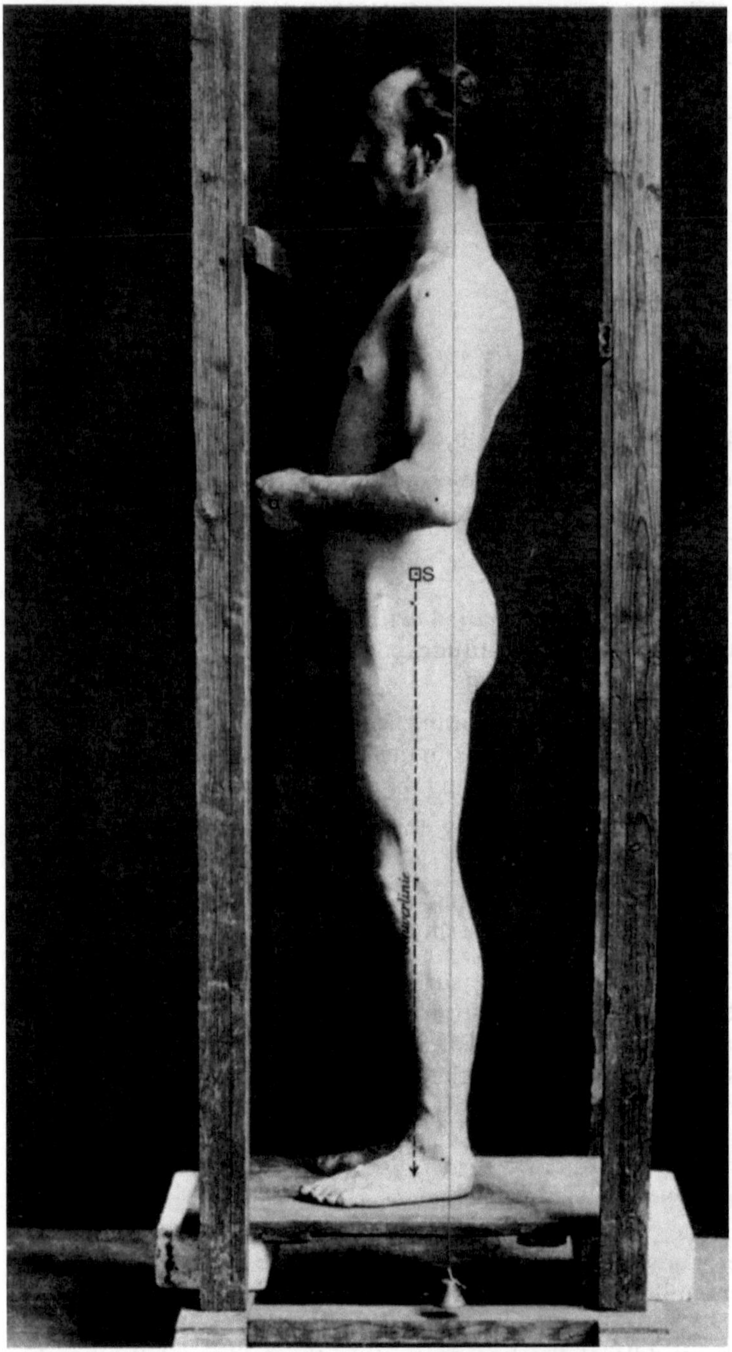

Fig. 8. Comfortable attitude, side view. · Projection of the centres of the joints; ⊡ projection of the centres of gravity of the head and hands; ⊡ *S* projection of the centre of gravity of the whole body

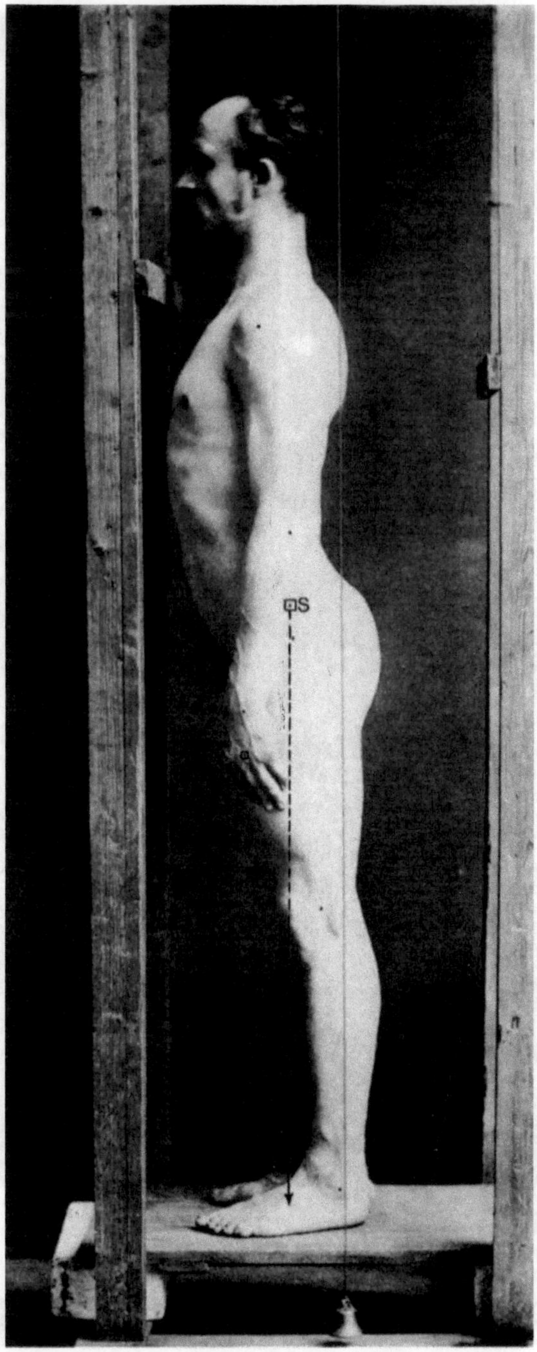

Fig. 9. Military attitude without regulation equipment, side view. · Projection of the centres of the joints; ⊡ projection of the centres of gravity of the head and hands; ⊡ *S* projection of the centre of gravity of the whole body.

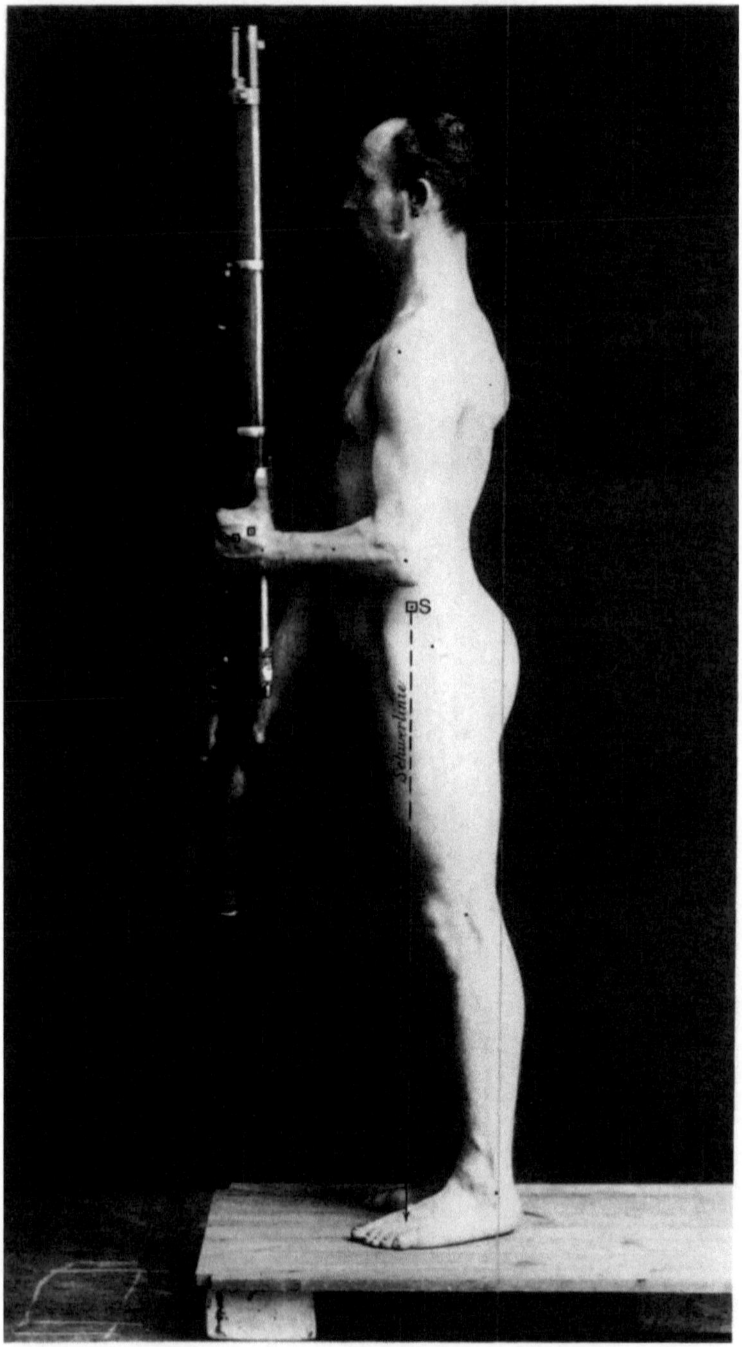

Fig. 10. "Present arms" military attitude without regulation equipment, side view. · Projection of the centres of the joints; ⊡ projection of the centres of gravity of the head, hands and rifle; ⊡ *S* projection of the centre of gravity of the whole body with rifle

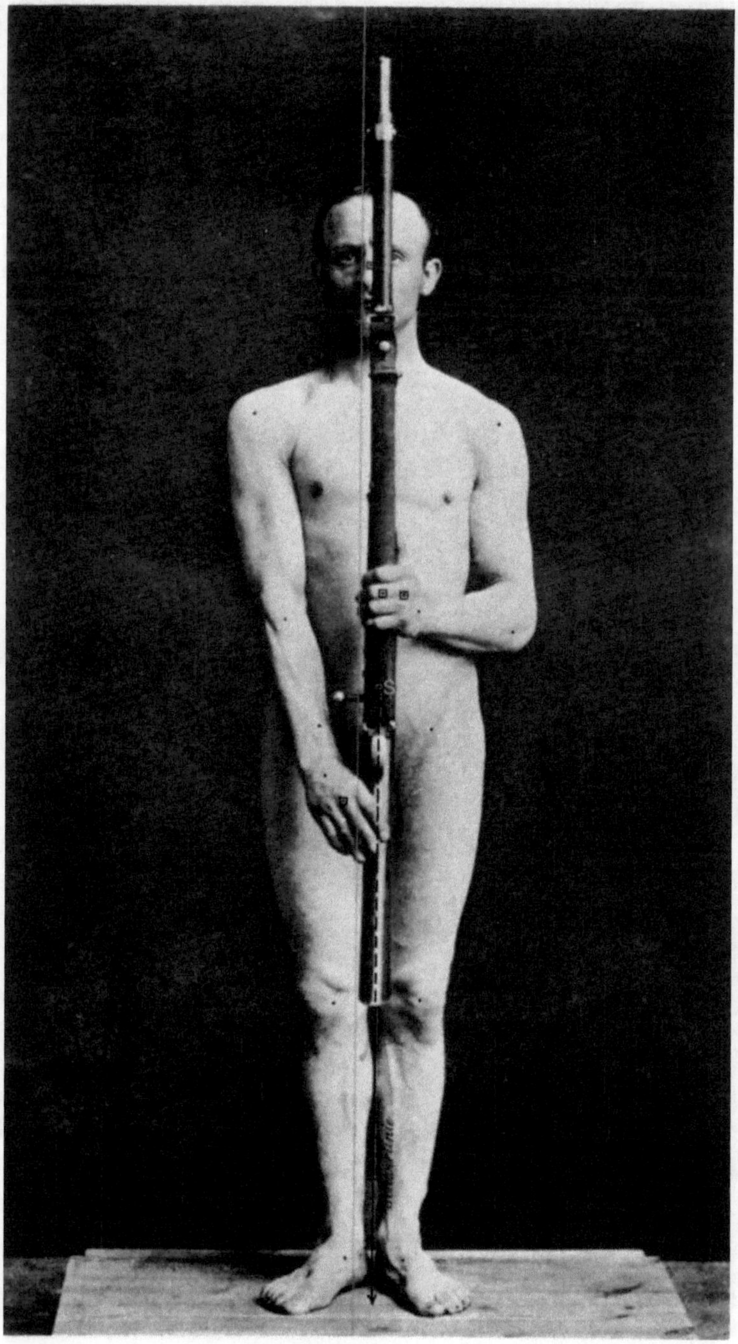

Fig. 11. "Present arms" military attitude without regulation equipment, front view. · Projection of the centres of the joints; ⊡ projection of the centres of gravity of the head, hands and rifle; ⊡ *S* projection of the centre of gravity of the whole body with rifle

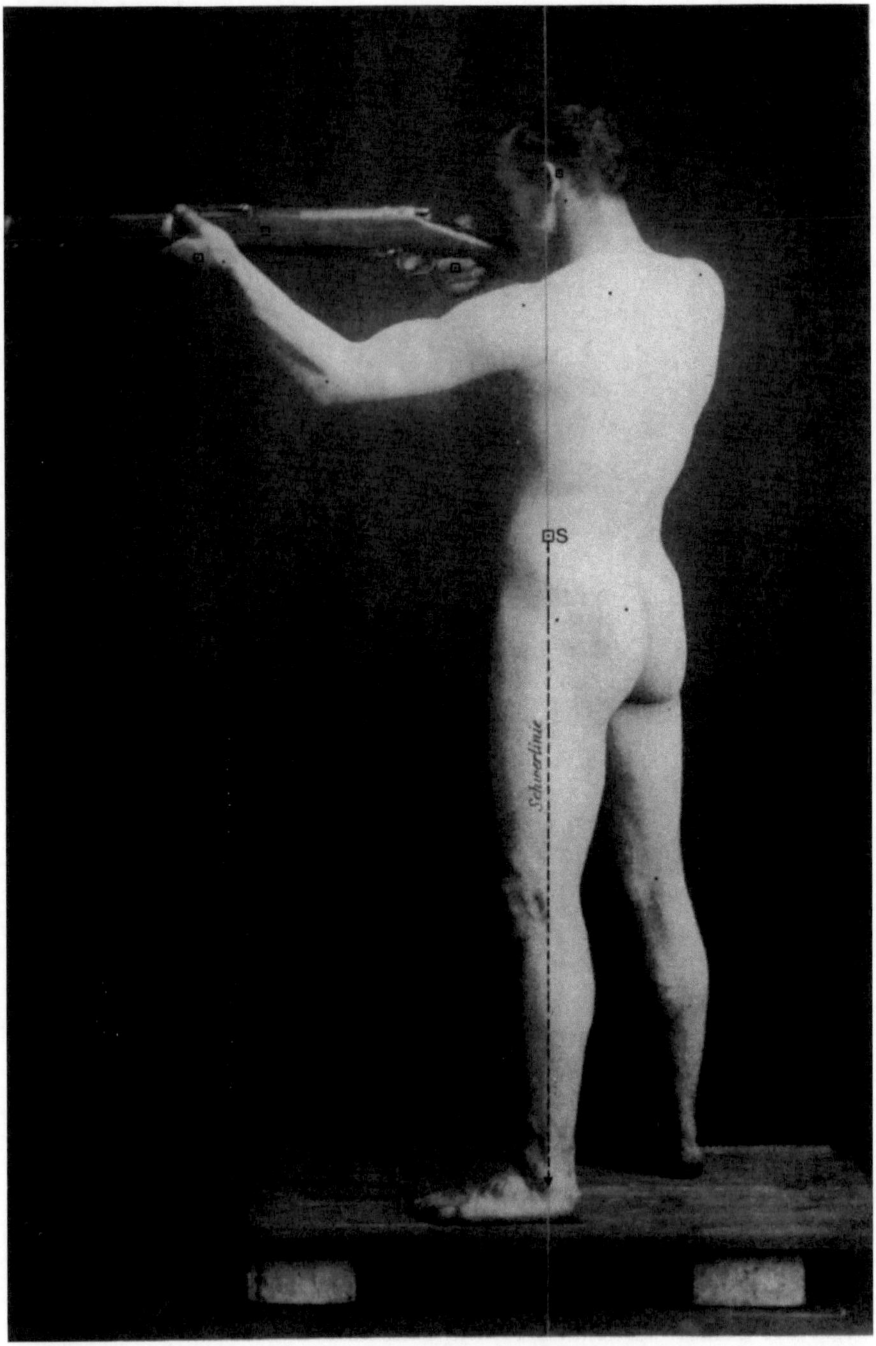

Fig. 12. Shooting attitude without regulation equipment, side view. · Projection of the centres of the joints; ⊡ projection of the centres of gravity of the head, hands and rifle; ⊡ *S* projection of the centre of gravity of the whole body with rifle

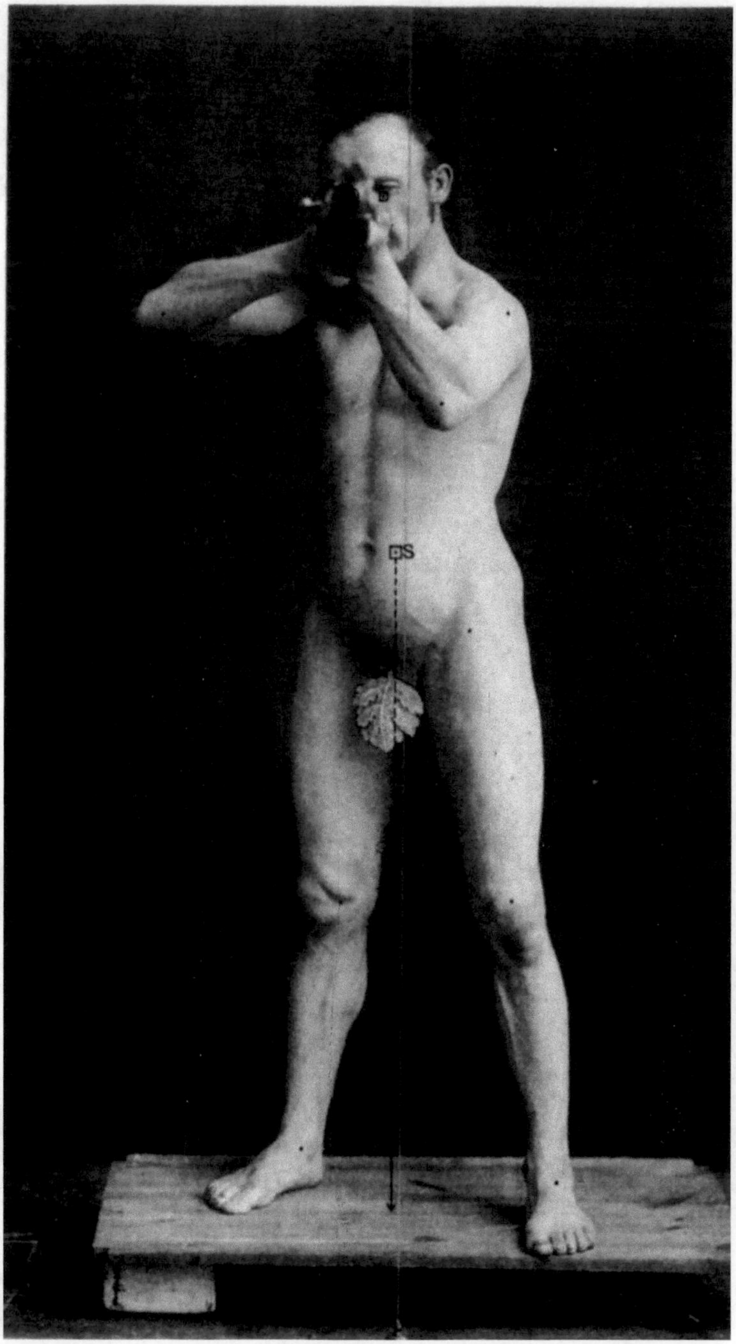

Fig. 13. Shooting attitude without regulation equipment, front view. · Projection of the centres of the joints; ⊡ projection of the centres of gravity of head, hands and rifle; □ *S* projection of the centre of gravity of the whole body with rifle

Fig. 14. Attitude with the rifle held at arm's length, without regulation equipment, side view. · Projection of the centres of the joints; ⊡ projection of the centres of gravity of the head, hands and rifle; ⊡ *S* projection of the centre of gravity of the whole body with rifle

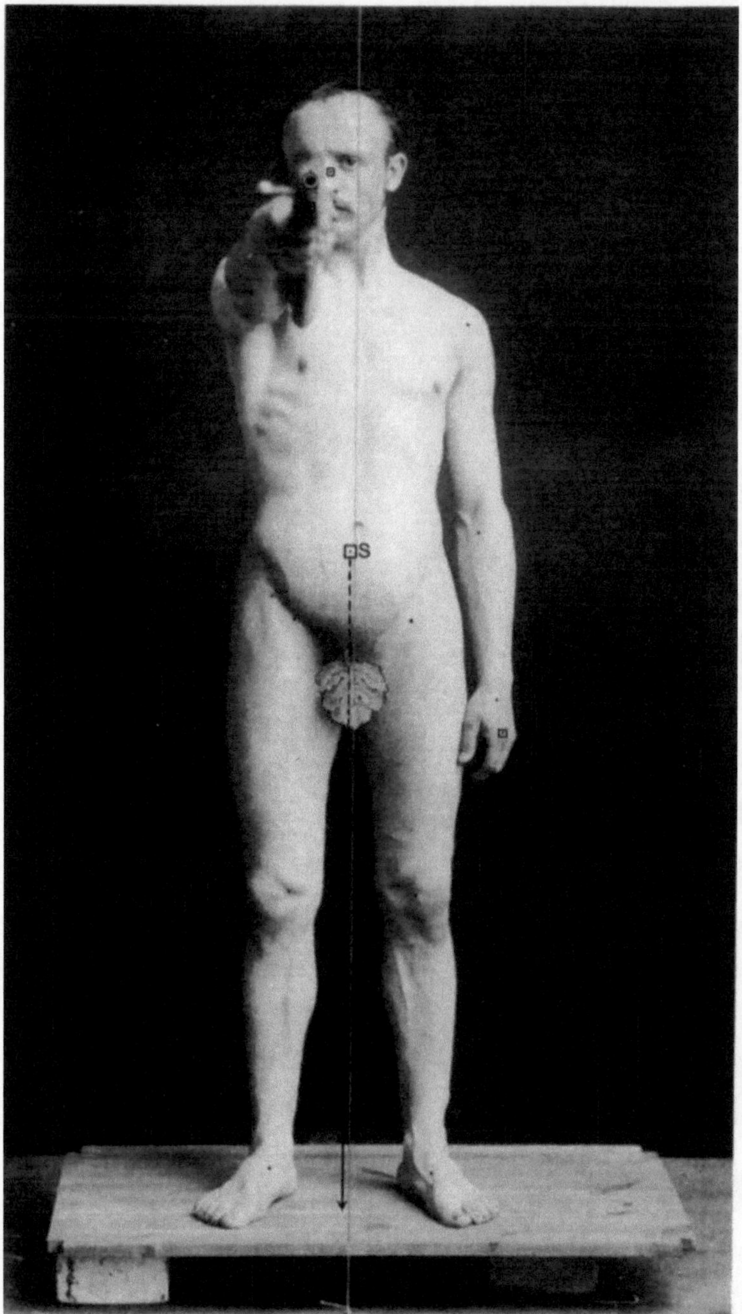

Fig. 15. Attitude with the rifle held at arm's length, without regulation equipment, front view. · Projection of the centres of the joints; ⊡ projection of the centres of gravity of the head, hands and rifle; ⊡ *S* projection of the centre of gravity of the whole body with rifle

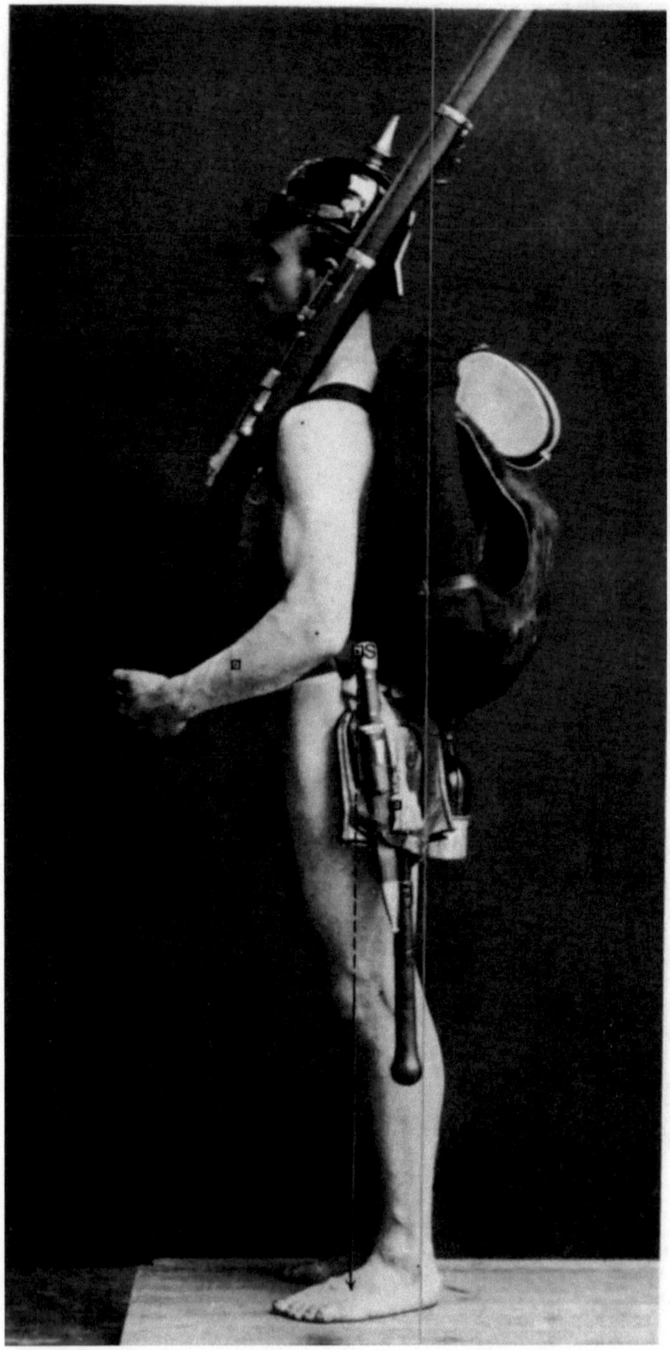

Fig. 16. Military attitude with regulation equipment at "shoulder arms", side view. · Projection of the centres of the joints; ⊡ projection of the centres of gravity of the head, hands and rifle; ⊡ *S* projection of the centre of gravity of the whole body with rifle

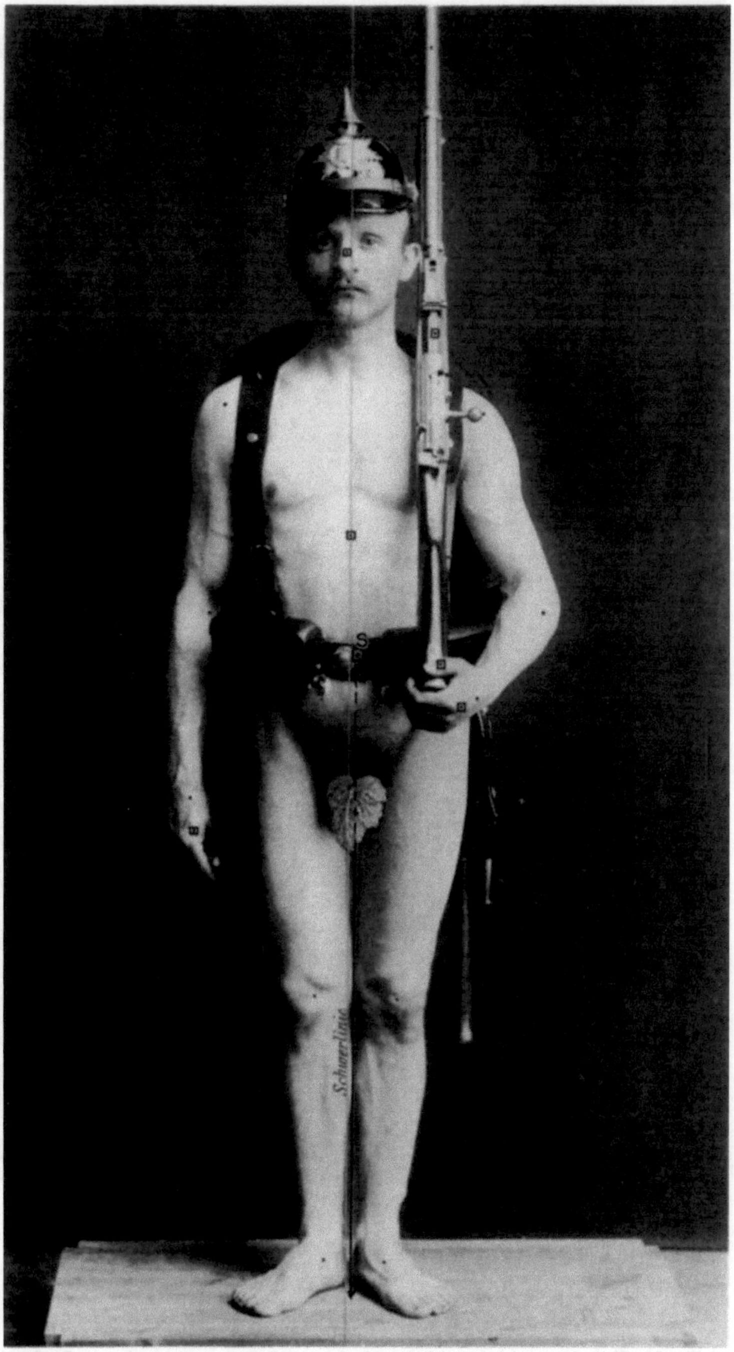

Fig. 17. Military attitude with regulation equipment at "shoulder arms", front view. · Projection of the centres of the joints; ⊡ projection of the centres of gravity of the head, hands and rifle; ⊡ *S* projection of the centre of gravity of the whole body with rifle

Fig. 18. Shooting attitude with regulation equipment, side view. · Projection of the centres of the joints; ⊡ projection of the centres of gravity of the head, hands and pieces of equipment; ⊡ *S* projection of the centre of gravity of the whole body with regulation equipment

Fig. 19. Shooting attitude with regulation equipment, front view. · Projection of the centres of the joints; ⊡ projection of the centres of gravity of the head, hands and pieces of equipment; ⊡ *S* projection of the centre of gravity of the whole body with the regulation equipment

For attitudes 1–3, only the X and Z co-ordinates can be obtained from one photograph. Since, however, we measured the distance between the symmetrical joints on the living subject, we also obtained the Y co-ordinates, assuming complete symmetry.

Attitudes of the Unloaded Body

Normal Attitude, Fig. 6

The co-ordinates of the centres of the joints are given in Table 24. From the data in Table 24, using the equations:

$$x_{i,k} = \varepsilon_k x_i + \varepsilon_i x_k$$

$$y_{i,k} = \varepsilon_k y_i + \varepsilon_i y_k$$

$$z_{i,k} = \varepsilon_k z_i + \varepsilon_i z_k$$

and the values of ε_i and ε_k given in Table 22, we calculated the co-ordinates of the centres of gravity given in Table 25 except those of the head and hands, which were measured directly from the pictures.

Using these co-ordinates and the weights given in Table 22, we then calculated the co-ordinates of the total centre of gravity (Table 26).

Table 24. Co-ordinates of the centres of the joints for the normal attitude

		x	y	z
Occipito-atlantal joint		0	0	152
Hip	r	0	+ 8.5	87
	l	0	− 8.5	87
Knee	r	0	+ 8.5	47
	l	0	− 8.5	47
Ankle	r	0	+ 8.5	6
	l	0	− 8.5	6
Posterior border of foot	r	− 4	+ 6	4
	l	− 4	− 6	4
Tip of foot	r	+20	+16	1.5
	l	+20	−16	1.5
Shoulder	r	0	+18	134
	l	0	−18	134
Elbow	r	0	+18	103
	l	0	−18	103
Wrist	r	0	+18	76
	l	+19	−11	108

Consequently:

$$x_0 = +\frac{35220}{58700}, \quad y_0 = +\frac{8786}{58700}, \quad z_0 = +\frac{5421379}{58700}$$

or: $x_0 = +0.6, \quad y_0 = +0.1, \quad z_0 = 92.4.$

Table 25. **Co-ordinates of the centres of gravity for the normal attitude**

		x	y	z
Head		0	0	156
Trunk		0	0	112.4
Thigh	r	0	+ 8.5	69.4
	l	0	− 8.5	69.4
Lower leg	r	0	+ 8.5	29.8
	l	0	− 8.5	29.8
Foot	r	+ 6.3	+ 10.3	3
	l	+ 6.3	− 10.3	3
Upper arm	r	0	+ 18	119.4
	l	0	− 18	119.4
Forearm	r	0	+ 18	91.7
	l	+ 8	− 15.1	105.1
Hand	r	0	+ 18	71
	l	+ 23	− 8	108.5

Table 26. **Calculation of the co-ordinates x_0, y_0, z_0 of the total centre of gravity in the normal attitude**

		$p_i x_i$	$p_i y_i$	$p_i z_i$
Head		0	0	645 840
Trunk		0	0	2 816 744
Thigh	r	0	+ 57 800	471 920
	l	0	− 57 800	471 920
Lower leg	r	0	− 26 265	92 082
	l	0	− 26 265	92 082
Foot	r	+ 6 615	+ 10 815	3 150
	l	+ 6 615	− 10 815	3 150
Upper arm	r	0	+ 35 640	236 412
	l	0	− 35 640	236 412
Forearm	r	0	+ 24 120	122 878
	l	+ 10 720	− 20 234	140 834
Hand	r	0	+ 8 820	34 790
	l	+ 11 270	− 3 920	53 165
Total		+ 35 220	+ 8 786	5 421 379

If the left forearm is extended and adopts a position symmetrical with the right, its co-ordinates become 0, − 18, 91.7, and those of the left hand become 0, − 18, 71. The final co-ordinates of the total centre of gravity for the normal attitude then become:

$$x_0 = + \frac{13230}{58700} = + 0.2$$

$$y_0 = 0$$

$$z_0 = \frac{5385048}{58700} = 91.7.$$

Raising the left forearm displaced the total centre of gravity 0.7 cm upwards, 0.4 cm forwards, and 0.1 cm to the right as mentioned on p. 27.

For a symmetrical normal attitude, the total centre of gravity thus lies 4.7 cm above the line connecting the centres of the hips whereas our direct measurement gave it as 4.5 cm above this line.

As mentioned above (p. 1), the support area of an erect human body is the surface delineated by the outlines of the feet and the double tangents to this outline. The stability of the attitude is at its greatest when the line of gravity intersects the middle of this surface. Figure 20 represents the projection of the

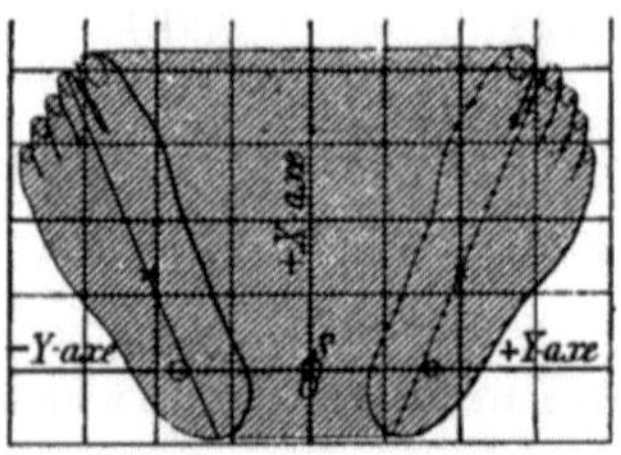

Fig. 20. Support area and its intersection ⊡ *S* by the line of gravity in the normal attitude, Fig. 6. The origin *O* of the co-ordinates is the projection of the middle of the line connecting the centres of the hips. ⊙ centre of the ankle joint, × centre of gravity of the foot. Scale: 1/10

feet on the plane *XY*. The intersection of the support area by the gravity line is indicated. It lies much closer to the posterior border than to the anterior. This attitude is thus more stable against a push from behind than against one from the front; the body is more prone to tip backwards than forwards. It thus appears that our normal attitude is not the safest. However, this attitude is very convenient for measurement and calculation. The centres of gravity of the body segments lie above each other in a vertical plane. As appears in Fig. 6, this attitude represents a reliable initial attitude.

In a remarkable work on the horizontal plane of the human skull which was based on measurements on an adult German skull and carefully carried out on living individuals, Schmidt noted that the plane through the external ear openings and through the lower border of the orbits is not physiologically horizontal[16]. This plane slopes upward ventrally, on average at an angle of $5\frac{1}{2}° - 5\frac{3}{4}°$ with the horizontal plane. This was confirmed by many observations. Schmidt designates as physiologically horizontal the horizontal plane when the head is

16 Schmidt (1876) Archiv für Anthropologie, vol. 9, p. 25

kept straight[17]. The head is straight when, in the erect attitude with a horizontally directed gaze, it rests on the spine with as little muscular effort as possible. Schmidt also found that the straight position of the head is adopted particularly by soldiers or former soldiers. He summarizes his work thus:

> The plane connecting the zygomatic bone above the external openings of the ears and the lower border of the orbits, the plane as defined by the anthropologists in Göttingen, constitutes the most horizontal plane that can be found; it is the one closest to the physiological horizontal plane. Of all the normal planes proposed, it is the most stable.

Without knowing the position of the centre of gravity of the head, a straight position of the head and a horizontal plane of the skull based on this position cannot be determined reliably. We can assume that, in life, the head is usually in unstable equilibrium since there is no ligamentum nuchae in man and the muscles of the neck cannot be considered to carry the head because nowhere in the human body do muscles constitute a permanent carrying apparatus. On this assumption, our constructed picture of Fig. 5 shows the head in the normal straight attitude. If a line is drawn connecting the middle of the opening of the ear and the lower border of the orbit, this line deviates from the horizontal by about 8°. This angle is very close to the values found by Schmidt. A series of measurements on macerated skulls, however, show that fairly large individual variations exist. It is thus better not to calculate in terms of degrees or minutes but to measure on a wider basis. We must leave the study of this question to the anthropologists and be satisfied with our approximation since only numerous and precise measurements can provide useful information.

The normal attitude is also convenient in determining the position of the centre of gravity of the body directly in vivo, without photographs. To this end, we must know the deviation of every body segment from the normal attitude, in direction and magnitude (degrees, as related to the centres of the two adjacent joints). In other words we must know the inclination of every body segment to the vertical. Thus, we must first calculate the co-ordinates of the centres of the joints and from them calculate the co-ordinates of the centres of gravity, as described above. Therefore, we must know how to calculate the co-ordinates of the centres of the joints adjacent to a body segment after flexion, knowing the co-ordinates before flexion (in normal attitude).

In Fig. 21, M_k and M_i represent the centres of the two joints adjacent to a body segment. M_i relates to the proximal joint, M_k to the distal. If the body segment M_iM_k is in the normal attitude, it is parallel to the axis Z. If the body segment is rotated about the joint centre M_k by an angle α, in relation to the normal attitude, in a plane forming an angle φ with the plane XZ, the joint centre M_i takes the position M_i'. α always designates the angle formed by the new position of the connecting line M_kM_i' with the positive direction of axis Z. The co-ordinates of point M_i' can then be calculated from those of point M_k, which remains fixed. a designates the length of the body segment, which is the distance between the centres of the two adjacent joints. M_kAM_i' is thus a right-angled tri-

[17] Schmidt, op. cit. p. 35

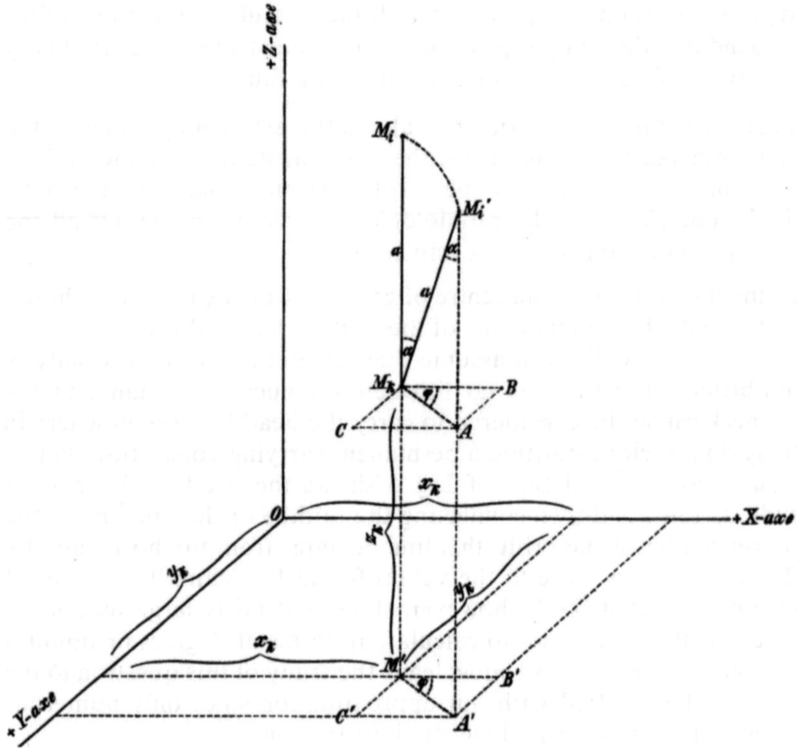

Fig. 21. Calculation of the co-ordinates of a joint centre M_i after flexion of the adjacent body segment $M_i M_k$, from its co-ordinates before flexion

angle, with the right angle at A, the acute angle α, and the hypotenuse a. Consequently, $M_k A = a \sin \alpha$. The segment $M_k A$ is the hypotenuse of the right-angled triangle ABM_k with the right angle at B and the acute angle φ. In this right-angled triangle, the two sides $M_k B$ and AB are the differences between x_i' and x_k and between y_i' and y_k. The co-ordinate z_i' is greater than z_k by the segment AM_i'. AM_i' being a side of the first right-angled triangle, we have:

$$x_i' = x_k + a \sin \alpha \cos \varphi$$
$$y_i' = y_k + a \sin \alpha \sin \varphi$$
$$z_i' = z_k + a \cos \alpha$$

The Comfortable Attitude, Fig. 8

The co-ordinates of the centres of the joints are given in Table 27, x_0, y_0, z_0 of the total centre of gravity in Table 29. From these data it follows that

$$x_0 = -\frac{30\,557}{58\,700}, \quad y_0 = +\frac{6734}{58\,700}, \quad z_0 = \frac{5\,387\,785}{58\,700}$$

or: $x_0 = -0.5, \quad y_0 = +0.1, \quad z_0 = 91.8.$

If the left forearm is extended so as to adopt an attitude symmetrical with that of the right, the three co-ordinates of the centre of gravity become:

– for the left forearm: −1.3, −18, 90.9
– for the left hand: 0, −18, 70.5

Table 27. Co-ordinates of the centres of the joints for the comfortable attitude

Joint		*x*	*y*	*z*
Occipito-atlantal		−1.5	0	150.5
Hip	r	0	+8.5	87
	l	0	−8.5	87
Knee	r	−1	+7	47
	l	−1	−7	47
Ankle	r	−5	+4.5	6
	l	−5	−4.5	6
Posterior border of foot	r	−9	+2	4
	l	−9	−2	4
Tip of foot	r	+15	+12	1.5
	l	+15	−12	1.5
Shoulder	r	−1	+18	133
	l	−1	−18	133
Elbow	r	−3	+18	102
	l	−3	−18	102
Wrist	r	+1	+18	75.5
	l	+17	−13	102.5

Table 28. Co-ordinates of the centres of gravity for the comfortable attitude

		x	*y*	*z*
Head		−1	0	154.5
Trunk		−0.6	0	111.8
Thigh	r	−0.4	+7.8	69.4
	l	−0.4	−7.8	69.4
Lower leg	r	−2.7	+6	29.8
	l	−2.7	−6	29.8
Foot	r	+1.3	+6.3	3
	l	+1.3	−6.3	3
Upper arm	r	−1.9	+18	118.4
	l	−1.9	−18	118.4
Forearm	r	−1.3	+18	90.9
	l	+5.4	−15.9	102.2
Hand	r	0	+18	70.5
	l	+20.5	−10	102

Table 29. Calculation of the co-ordinates x_0, y_0, z_0 of the total centre of gravity in the comfortable attitude

		$p_i x_i$	$p_i y_i$	$p_i z_i$
Head		− 4 140	0	639 630
Trunk		− 15 036	0	2 801 708
Thigh	r	− 2 720	+ 53 040	471 920
	l	− 2 720	− 53 040	471 920
Lower leg	r	− 8 343	+ 18 540	92 082
	l	− 8 343	− 18 540	92 082
Foot	r	+ 1 365	+ 6 615	3 150
	l	+ 1 365	− 6 615	3 150
Upper arm	r	− 3 762	+ 35 640	234 432
	l	− 3 762	− 35 640	234 432
Forearm	r	− 1 742	+ 24 120	121 806
	l	+ 7 236	− 21 306	136 948
Hand	r	0	+ 8 820	34 545
	l	+ 10 045	− 4 900	49 980
Total		− 30 557	+ 6 734	5 387 785

Consequently, the co-ordinates of the total centre of gravity in the comfortable attitude become:

$$x_0 = -\frac{49\,580}{58\,700} = -0.8$$

$$y_0 = 0$$

$$z_0 = \frac{5\,357\,208}{58\,700} = 91.3.$$

Raising the forearm in this case displaced the centre of gravity 0.3 cm forward, 0.1 cm to the right, and 0.5 cm upward. In the comfortable attitude, if the two sides are symmetrical, the total centre of gravity thus lies 7.3 cm higher than and 0.8 cm behind the centres of the hip, above the sockets.

Figure 22 represents the projection of the feet on the plane XY and the intersection of the support area by the line of gravity. In this case, the line is 4 cm more anterior than in the normal attitude. This results in better stability of the attitude.

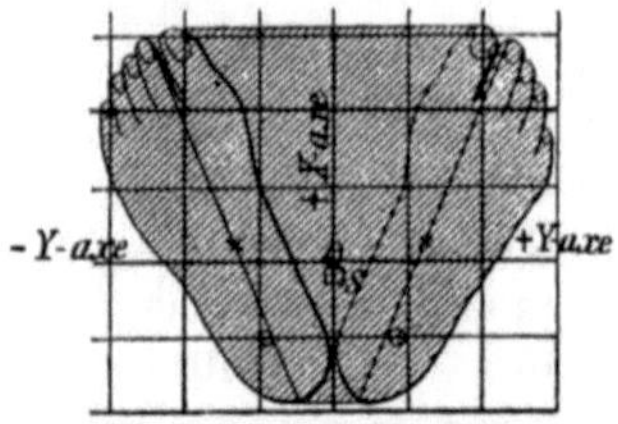

Fig. 22. Support area and its intersection ⊡ S by the line of gravity in the comfortable attitude, Fig. 8. The origin O of the co-ordinates is the projection of the middle of the line connecting the centres of the hips. ⊙ centre of the ankle joint, × centre of gravity of the foot. Scale: 1/10

The Military Attitude, Fig. 9

Co-ordinates of the centres of the joints are given in Table 30, those of the centres of gravity in Table 31, and from these co-ordinates the co-ordinates x_0, y_0, z_0 of the total centre of gravity can be calculated (Table 32).

Consequently:

$$x_0 = +\frac{30611}{58700}, \quad y_0 = \frac{0}{58700}, \quad z_0 = \frac{5381005}{58700}$$

or: $x_0 = +0.5, \quad y_0 = 0, \quad z_0 = 91.7.$

In this case the left forearm was not raised but pushed a little forward as required by the military attitude. This made the points visible. The two arms were in a symmetrical position. The co-ordinates of the left forearm were then: +0.3, −18, 91.4; for the left hand they were: 0, −18, 70. Consequently, the co-ordinates of the total centre of gravity in the military attitude were:

$$x_0 = +\frac{22782}{58700} = +0.4$$

$$y_0 = 0$$

$$z_0 = \frac{5380492}{58700} = 91.7$$

Table 30. Co-ordinates of the centres of the joints in the military attitude (co-ordinates taken from photographs)

Joint		x	y	z
Occipito-atlantal		+ 5	0	152
Hip	r	0	+ 8.5	87
	l	0	− 8.5	87
Knee	r	− 4.5	+ 6	47
	l	− 4.5	− 6	47
Ankle	r	− 7	+ 4.5	6
	l	− 7	− 4.5	6
Posterior border of foot	r	− 11	+ 2	4
	l	− 11	− 2	4
Tip of foot	r	+ 13	+ 12	1.5
	l	+ 13	− 12	1.5
Shoulder	r	+ 4.5	+ 18	133
	l	+ 4.5	− 18	133
Elbow	r	+ 0.5	+ 18	102.5
	l	+ 0.5	− 18	102.5
Wrist	r	0	+ 18	76
	l	+ 7.5	− 18	76.5

Table 31. Co-ordinates of the centres of gravity for the military attitude

		x	y	z
Head		+6	0	156
Trunk		+2	0	112.4
Thigh	r	−2	+ 7.4	69.4
	l	−2	− 7.4	69.4
Lower leg	r	−5.6	+ 5.4	29.8
	l	−5.6	− 5.4	29.8
Foot	r	−0.7	+ 6.3	3
	l	−0.7	− 6.3	3
Upper arm	r	+2.6	+18	118.7
	l	+2.6	−18	118.7
Forearm	r	+0.3	+18	91.4
	l	+3.4	−18	91.6
Hand	r	0	+18	70
	l	+7.5	−18	70.5

Table 32. Calculation of the co-ordinates x_0, y_0, z_0 of the total centre of gravity in the military attitude

		$p_i x_i$	$p_i y_i$	$p_i z_i$
Head		+24 840	0	645 840
Trunk		+50 120	0	2 816 744
Thigh	r	− 13 600	+50 320	471 920
	l	− 13 600	−50 320	471 920
Lower leg	r	− 17 304	+16 686	92 082
	l	− 17 304	−16 686	92 082
Foot	r	− 735	+ 6 615	3 150
	l	− 735	− 6 615	3 150
Upper arm	r	+ 5 148	+35 640	235 026
	l	+ 5 148	−35 640	235 026
Forearm	r	+ 402	+24 120	122 476
	l	+ 4 556	−24 120	122 744
Hand	r	0	+ 8 820	34 300
	l	+ 3 675	− 8 820	34 545
Total		+30 611	0	5 381 005

This incorrect attitude of the left forearm, therefore, hardly changed the position of the total centre of gravity, which was displaced 1 mm anteriorly but kept its initial height.

In the military attitude, the total centre of gravity remains at the same height as in the normal attitude; it lies 4.7 cm higher than the centres of the hips. It is displaced only 0.4 cm forward in the body so that it remains vertically above

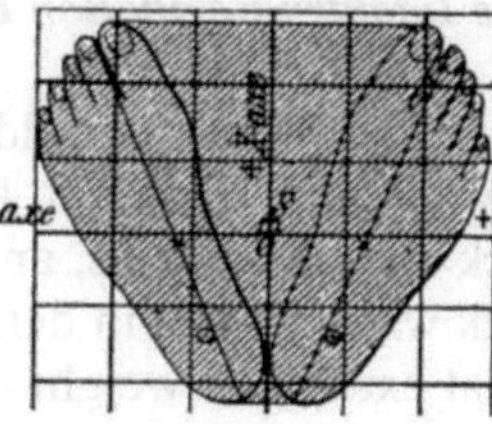

Fig. 23. Support area and its intersection ⊡ *S* by the line of gravity in the military attitude, Fig. 9. The origin *O* of the co-ordinates is the projection of the middle of the line connecting the centres of the hips. ⊙ Centre of the ankle joint, × centre of gravity of the foot. Scale: 1/10

the hip joints. However, the line of gravity intersects the support area much more anteriorly than in the normal attitude, because of the forward inclination of the subject.

Figure 23 illustrates the position of the line of gravity in the support area. The line of gravity intersects the support area 7.2 cm in front of the ankle joints, thus closer to the metatarsal heads than to the heels. In the normal attitude, stability is lower against a push from the front; in the military attitude it is lower against a push from behind: the body tends to fall forward rather than backward. As far as stability is concerned, the comfortable attitude (Fig. 8) is intermediary between the two other attitudes; it offers almost the same stability against a push from the front as from behind.

The unfavourable position of the line of gravity in the military attitude results in considerable demand on the muscles, which leads to fatigue when this attitude is maintained for some time. Therefore, this attitude is not recommended for standing for a long time. It seems to be appropriate only in facilitating the first step in walking.

As mentioned on p. 5, Meyer locates the centre of gravity for the erect stable attitude, which he designates as military, in the second sacral vertebra or immediately above, in the sacral canal, such that the line of gravity lies 5 cm behind the hip axis. This corresponds neither to our military attitude nor to our stable comfortable attitude. Meyer's attitude constitutes an enforced extreme attitude with excessive arching of the trunk backwards. This attitude would be extremely uncomfortable and fatiguing.

Attitudes of the Loaded Body

The army put at our disposal a complete set of M/87 regulation equipment[19]. For our studies, the subject did not wear the uniform, which would have hidden the outlines of the body and the joint marks. Accordingly, we did not consider the centre of gravity of the helmet in our calculations. We thus deliberately discarded part of the soldier's load. However, since the clothes almost completely cover the body, their removal can hardly displace the centre of gravity. The heavy boots are partly counterbalanced by the weight of the helmet. Similarly, the belt was considered as part of the uniform and not included in the calculations.

---

[19] We would like to thank General Major Walde and Major Meissner of the 8th Infantry Regiment, Prinz Johann Georg Nr. 107, for their co-operation and kind support of our work, without which this research could not have been carried out.

The Infantry Regulation Equipment M/87

For our studies it would be sufficient just to give the weights of the rifle, the bayonet and spade, the filled cartridge pouches, the filled knapsack, the haversack and its contents, and the water bottle. However, the contents of the knapsack will be given in detail (Table 33). This will make it possible to calculate to what extent the weight of the knapsack can be increased by additional objects and how omitting particular items can decrease the load and change the position of the centre of gravity.

Table 33. Data on the weight of the infantry regulation equipment M/87

Equipment	Weight (kg)
1 knapsack	
1 bag	
1 set of straps for the knapsack	
1 back cartridge pouch	
40 live cartridges	
1 shirt	
1 pair of underpants	
1 pair of white trousers	
2 pairs of footcloths	
2 pairs of socks	
3 handkerchiefs	
1 paybook	
1 prayer book	
1 clothes brush	12.250
1 shining brush	
1 cleaning brush	
1 polishing brush	
1 sewing kit	
3 meat tins	
4 tins for rifle oil, dubbin, boot polish	
3 vegetable tins	
3 biscuit rations	
3 bags (salt, coffee, rice)	
1 cap	
1 overcoat with 2 straps	
1 canteen with 2 straps	
1 belt with buckle and pouch	
2 front cartridge pouches	
60 live cartridges	
1 bayonet	7.300
1 spade with cover	
1 haversack with contents	
1 water bottle	
1 rifle	
1 rifle strap	4.700
Rifle sight	
Total weight of the equipment	24.250

Table 34. Weights of the different pieces of equipment in the study

Equipment	Weight (g)
1. Knapsack with back cartridge pouch (with 40 live cartridges)	12 250
2. Two front cartridge pouches each with 30 live cartridges (2×1 590 g)	3 180
3. Bayonet and spade	1 570
4. Full haversack with water bottle	1 570
5. Rifle	4 700
Total 1–5	23 270
6. Helmet	525
7. Belt with buckle	305
Total 6 and 7	830
Total 1–7	24 100

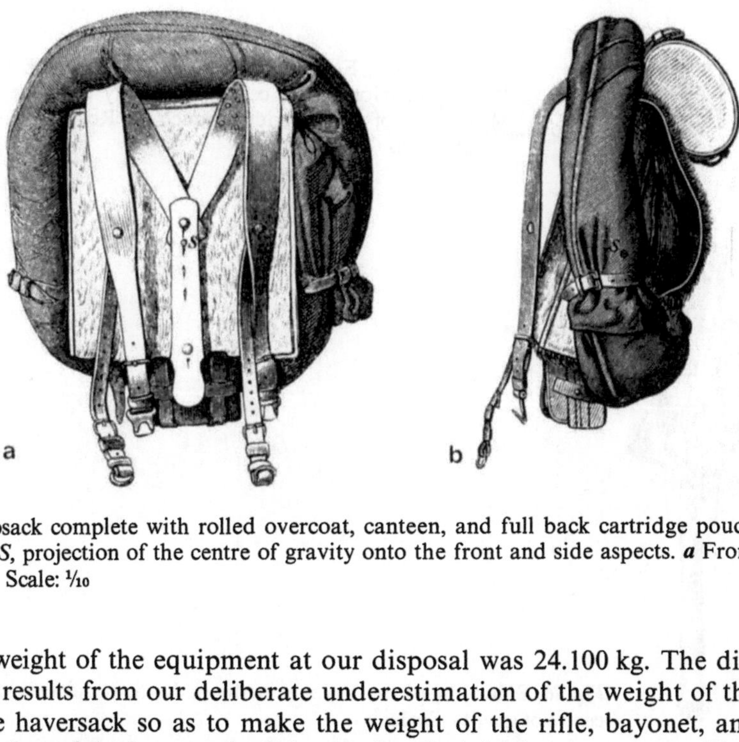

Fig. 24a, b. Knapsack complete with rolled overcoat, canteen, and full back cartridge pouch (live cartridges). *S*, projection of the centre of gravity onto the front and side aspects. ***a*** Front view; ***b*** side view. Scale: 1/10

The actual weight of the equipment at our disposal was 24.100 kg. The difference of 150 results from our deliberate underestimation of the weight of the contents of the haversack so as to make the weight of the rifle, bayonet, and spade equal to that of the haversack and water bottle (which can vary considerably anyway). This was done to facilitate the calculation (Table 34).

The positions of the centres of gravity of the various pieces of equipment were determined by suspending each piece of equipment in three different ways.

– The centre of gravity of the full knapsack (Fig. 24) with the full back cartridge pouch was 155 mm above the lower border of the knapsack (note, not the cartridge pouch) and 185 mm below the upper border of the knapsack (not the overcoat), exactly in the median plane dividing the knapsack into a right and left half, 70 mm behind the frame of its anterior surface and, since the packing bulged anteriorly, 100 mm behind the centre of the bulging.

- The centre of gravity of a front cartridge pouch (Fig. 25) was 50 mm from its upper border, 34 mm from its lower border, and 22 mm from the mid-anterior surface.
- The centre of gravity of the bayonet, spade, and cover (Fig. 26) was 265 mm below the upper border of the cover, somewhat anterior.
- The common centre of gravity of the haversack with the water bottle changed position depending on their contents. It was assumed to be symmetrical with the common centre of gravity of the bayonet and spade.
- The centre of gravity of the M/71.84 rifle without the muzzle cover and without the sight was in a transverse plane of the rifle, 20 mm from the lower border of the sight, 15 mm below the upper surface of the shaft, and 26 mm above its lower surface (Fig. 27).

Fig. 25a, b. Front cartridge pouch filled with live cartridges. *S*, projection of the centre of gravity onto the rear and side aspects of the cartridge pouch. ***a*** Rear view; ***b*** side view. Scale: 1/10

Fig. 26. Side view of the bayonet with spade. *S*, projection of the common centre of gravity. Scale: 1/10

Fig. 27. Side view of the M/71.84 rifle. *S*, projection of the centre of gravity onto the surface of the shaft. Scale: 1/10

The "Present Arms" Military Attitude, Figs. 10, 11

The co-ordinates of the centres of the joints are given in Table 35, the co-ordinates of the centres of gravity are given in Table 36. The co-ordinates of the centre of gravity of the rifle were determined directly from the photographs, as were those of the head and hands. The calculation of the co-ordinates x_0, y_0, z_0 of the total centre of gravity is given in Table 37. Consequently:

$$x_0 = +\frac{186455}{63409}, \quad y_0 = +\frac{8354}{63400}, \quad z_0 = \frac{5894666}{63400}$$

$$\text{or: } x_0 = +2.9, \quad y_0 = +0.1, \quad z_0 = 93.0.$$

Figure 28 represents the support area and its intersection S by the line of gravity for the "presenting arms" military attitude. The line of gravity in this case lies much closer to the tip of the foot than to its posterior border. Therefore, this attitude is extremely unstable and tends to make the subject overbalance

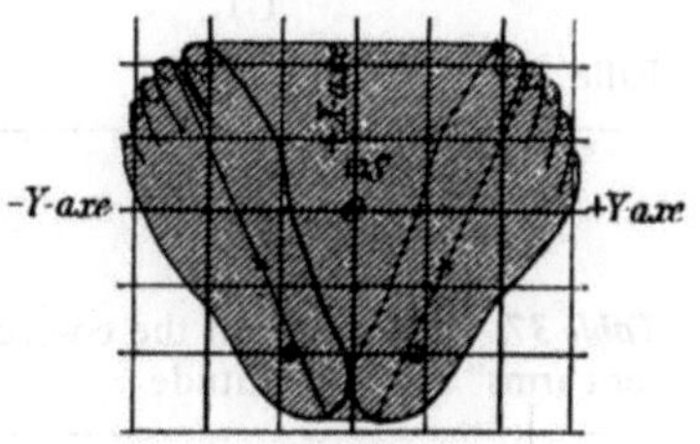

Fig. 28. Support area and its intersection. ⊡ S by the line of gravity in the "present arms" attitude, Figs. 10, 11. The origin O of the co-ordinates is the projection of the middle of the line connecting the centres of the hips. ⊙ Centre of the ankle joint, × centre of gravity of the foot. Scale: 1/10

Table 35. Co-ordinates of the centres of the joints for the "present arms" military attitude

Joint		x	y	z
Occipito-atlantal		+ 3	+ 1.5	151
Hip	r	0	+ 8.5	87
	l	0	− 8.5	87
Knee	r	− 5.5	+ 6	47
	l	− 5.5	− 6	47
Ankle	r	− 10	+ 4.5	6
	l	− 10	− 4.5	6
Posterior border of foot	r	− 14	+ 2	4
	l	− 14	− 2	4
Tip of foot	r	+ 10	+ 12	1.5
	l	+ 10	− 12	1.5
Shoulder	r	+ 5.5	+ 18	133
	l	+ 5.5	− 18	131
Elbow	r	+ 15	+ 13.5	103
	l	+ 4	− 20	101
Wrist	r	+ 23	+ 7.5	81
	l	+ 25.5	− 7.5	102

Table 36. Co-ordinates of the centres of gravity for the "present arms" military attitude

		x	y	z
Head		+ 4	+ 1.5	155
Trunk		+ 1.2	+ 0.6	122
Thigh	r	− 2.4	+ 7.4	69.4
	l	− 2.4	− 7.4	69.4
Lower leg	r	− 7.4	+ 5.4	29.8
	l	− 7.4	− 5.4	29.8
Foot	r	− 3.7	+ 6.3	3
	l	− 3.7	− 6.3	3
Upper arm	r	+ 10	+ 15.9	118.9
	l	+ 4.8	− 18.9	116.9
Forearm	r	+ 18.4	+ 11	93.8
	l	+ 13	− 14.8	101.4
Hand	r	+ 27	+ 4.5	77
	l	+ 29.5	− 3.5	103
Rifle		+ 27	− 0.5	104.5

Table 37. Calculation of the co-ordinates x_0, y_0, z_0 of the total centre of gravity in the "present arms" military attitude

		$p_i x_i$	$p_i y_i$	$p_i z_i$
Head		+ 16 560	+ 6 210	645 840
Trunk		+ 30 072	+ 15 036	2 806 720
Thigh	r	− 16 320	+ 50 320	471 920
	l	− 16 320	− 50 320	471 920
Lower leg	r	− 22 866	+ 16 686	92 082
	l	− 22 866	− 16 686	92 082
Foot	r	− 3 885	+ 6 615	3 150
	l	− 3 885	− 6 615	3 150
Upper arm	r	+ 19 800	+ 31 482	235 422
	l	+ 9 504	− 37 422	231 462
Forearm	r	+ 24 656	+ 14 740	125 692
	l	+ 17 420	− 19 832	135 876
Hand	r	+ 13 230	+ 2 205	37 730
	l	+ 14 455	− 1 715	50 470
Rifle		+ 126 900	− 2 350	491 150
Total		+ 186 455	+ 8 354	5 894 666

forwards. The centre of gravity lies 12.9 cm in front of the ankle joints, 2.9 cm in front of and 6 cm above the hip joints. Because of the presence of the rifle and the consequent attitude of the left forearm, the centre of gravity is 1.3 cm higher than in the military attitude of Fig. 9. To determine how far the rifle alone displaces the centre of gravity forwards and upwards, we must calculate the co-ordinates of the centre of gravity in this attitude with the position of the arms unchanged but without rifle. Therefore:

$$x_0 = +\frac{59555}{58700} = +1.0$$

$$y_0 = +\frac{10704}{58700} = +0.2$$

$$z_0 = \frac{5403516}{58700} = 92.1\,.$$

The rifle alone thus displaces the centre of gravity only 0.9 cm upwards and 1.9 cm forwards. The arms held forwards already displace the centre of gravity upwards and forwards, compared with the pure military attitude (Fig. 9). Moreover, it must be noticed that in presenting arms, the body leans forwards under the effect of the load of the rifle held forwards, more than in the military attitude without the weapon. This alone causes an anterior displacement of the centre of gravity.

The Shooting Attitude Without the Knapsack, Figs. 12, 13

The co-ordinates of the centres of the joints are given in Table 38, the co-ordinates of the centres of gravity are given in Table 39. The calculation of the co-ordinates x_0, y_0, z_0 of the total centre of gravity is given in Table 40. It follows that:

$$x_0 = +\frac{406461}{63400}, \quad y_0 = +\frac{169891}{63400}, \quad z_0 = \frac{6197950}{63400}$$

$$\text{or: } x_0 = +6.4, \quad y_0 = +2.7, \quad z_0 = 97.8\,.$$

The centre of gravity in this case lies 11.8 cm above the line connecting the centres of the hips, 7.1 cm higher than in the normal attitude. It lies above the sacral promontory, approximately at the level of the disc between the fourth and fifth lumbar vertebrae, about 5.5 cm in front of the latter, about 1 cm to the left of the median plane, and 7 cm in front of the middle of the line connecting the centres of the hips.

Figure 29 represents the intersection of the support area by the line of gravity. This attitude of the feet provides a large support area. The line of gravity passes closer to the anterior border of the support area than to its posterior border; this ensures stability against the recoil of the rifle. The position of the right foot backwards and sideways provides another advantage: the distance of the line of gravity from the posterior border of the foot is further increased in the direction of the negative axis X, i.e. in the direction of the recoil of the rifle.

Table 38. Co-ordinates of the centres of the joints for the shooting attitude without the knapsack

Joint		x	y	z
Occipito-atlantal		+ 3.5	+ 3.5	147
Hip	r	− 5	+ 7.5	86
	l	+ 5	− 7.5	86
Knee	r	− 9.5	+ 11	46.5
	l	+ 8	− 13	46.5
Ankle	r	− 11	+ 16	6
	l	+ 7	− 19	6
Posterior border of foot	r	− 14.5	+ 12.5	4
	l	+ 2	− 18.5	4
Tip of foot	r	+ 5	+ 29.5	1.5
	l	+ 28	− 18.5	1.5
Shoulder	r	− 16	+ 15	137
	l	+ 9.5	− 14	132.5
Elbow	r	− 2.5	+ 37	133
	l	+ 38.5	− 3.5	120
Wrist	r	+ 16.5	+ 15.5	141.5
	l	+ 54	+ 8	141

Table 39. Co-ordinates of the centres of gravity in the shooting attitude without the knapsack

		x	y	z
Head		+ 4.5	+ 5	151
Trunk		+ 1.4	+ 1.4	109.8
Thigh	r	− 7	+ 9	68.6
	l	+ 6.3	− 9.9	68.6
Lower leg	r	− 10.1	+ 13.1	29.5
	l	+ 7.6	− 15.5	29.5
Foot	r	− 6.1	+ 19.8	3
	l	+ 13.2	− 18.5	3
Upper arm	r	− 9.7	+ 25.3	135.1
	l	+ 23.1	− 9.1	126.6
Forearm	r	+ 5.5	+ 28	136.6
	l	+ 45	− 1.3	128.8
Hand	r	+ 20	+ 12.5	142.5
	l	+ 57.5	+ 8.5	142.5
Rifle		+ 48	+ 9.5	146

Table 40. Calculation of the co-ordinates x_0, y_0, z_0 of the total centre of gravity in the shooting attitude without the knapsack

		$p_i x_i$	$p_i y_i$	$p_i z_i$
Head		+ 18 630	+ 20 700	625 140
Trunk		+ 35 084	+ 35 084	2 751 588
Thigh	r	− 47 600	+ 61 200	466 480
	l	+ 42 840	− 67 320	466 480
Lower leg	r	− 31 209	+ 40 479	91 155
	l	+ 23 484	− 47 895	91 155
Foot	r	− 6 405	+ 20 790	3 150
	l	+ 13 860	− 19 425	3 150
Upper arm	r	− 19 206	+ 50 094	267 498
	l	+ 45 738	− 18 018	250 668
Forearm	r	+ 7 370	+ 37 520	183 044
	l	+ 60 300	+ 1 742	172 592
Hand	r	+ 9 800	+ 6 125	69 825
	l	+ 28 175	+ 4 165	69 825
Rifle		+225 600	+ 44 650	686 200
Total		+406 461	+169 891	6 197 950

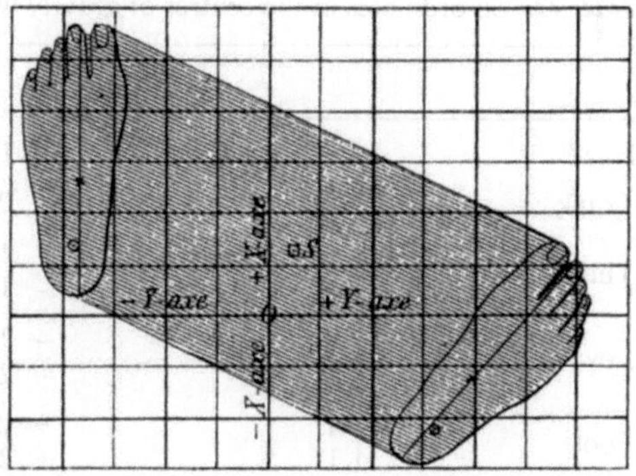

Fig. 29. Support area and its intersection ⊡ *S* by the line of gravity in the shooting attitude without regulation equipment, Figs. 12, 13. The origin *O* of the co-ordinates is the projection of the middle of the line connecting the centres of the hips. ⊙ Centre of the ankle joint, × centre of gravity of the foot. Scale: ⅒

Attitude with the Rifle Held Forward at Arm's Length, Figs. 14, 15

The co-ordinates of the centres of the joints are given in Table 41, the co-ordinates of the centres of gravity are given in Table 42. The calculation of the co-ordinates x_0, y_0, z_0 of the total centre of gravity is given in Table 43.

Consequently:

$$x_0 = +\frac{485\,510}{63\,400}, \quad y_0 = +\frac{60\,095}{63\,400}, \quad z_0 = \frac{6\,152\,147}{63\,400}$$

or: $x_0 = +7.7, \quad y_0 = +0.9, \quad z_0 = 97.0.$

Table 41. Co-ordinates of the centres of the joints with the rifle held forwards at arm's length

Joint		x	y	z
Occipito-atlantal		0	+ 3	150
Hip	r	0	+ 8.5	87
	l	0	− 8.5	87
Knee	r	− 2	+ 9	47
	l	− 2	− 9.5	47
Ankle	r	− 4	+ 11	6
	l	− 4	− 12.5	6
Posterior border of foot	r	− 8.5	+ 9	4
	l	− 8.5	− 10.5	4
Tip of foot	r	+ 16.5	+ 16	1.5
	l	+ 16.5	− 17.5	1.5
Shoulder	r	+ 2.5	+ 17.5	137
	l	− 5	− 16.5	131
Elbow	r	+ 29.5	+ 15.5	131
	l	− 5	− 18	100
Wrist	r	+ 53.5	+ 12.5	140.5
	l	+ 13.5	− 21	74

Table 42. Co-ordinates of the centres of gravity with the rifle held forwards at arm's length

		x	y	z
Head		+ 1	+ 4	154
Trunk		0	+ 1.2	111.6
Thigh	r	− 1.1	+ 8.7	69.4
	l	− 1.1	− 8.9	69.4
Lower leg	r	− 2.8	+ 9.8	29.8
	l	− 2.8	− 10.8	29.8
Foot	r	+ 2.3	+ 12	3
	l	+ 2.3	− 13.5	3
Upper arm	r	+ 15.2	+ 16.6	134.2
	l	− 5	− 17.2	116.4
Forearm	r	+ 39.6	+ 14.2	135
	l	− 1.4	− 19.3	89.1
Hand	r	+ 57	+ 11	140.5
	l	+ 6	− 22	69
Rifle		+ 86.5	+ 7	143.5

Table 43. Calculation of the co-ordinates x_0, y_0, z_0 of the total centre of gravity with the rifle held forewards at arm's length

		$p_i x_i$	$p_i y_i$	$p_i z_i$
Head		+ 4 140	+16 560	637 560
Trunk		0	+30 072	2 796 696
Thigh	r	− 7 480	+59 160	471 920
	l	− 7 480	−60 520	471 920
Lower leg	r	− 8 652	+30 282	92 082
	l	− 8 652	−33 372	92 082
Foot	r	+ 2 415	+12 600	3 150
	l	+ 2 415	−14 175	3 150
Upper arm	r	+ 30 096	+32 868	265 716
	l	− 9 900	−34 056	230 472
Forearm	r	+ 53 064	+19 028	180 900
	l	− 1 876	−25 862	119 394
Hand	r	+ 27 930	+ 5 390	68 845
	l	+ 2 940	−10 780	33 810
Rifle		+406 550	+32 900	674 450
Total		+485 550	+60 095	+6 152 147

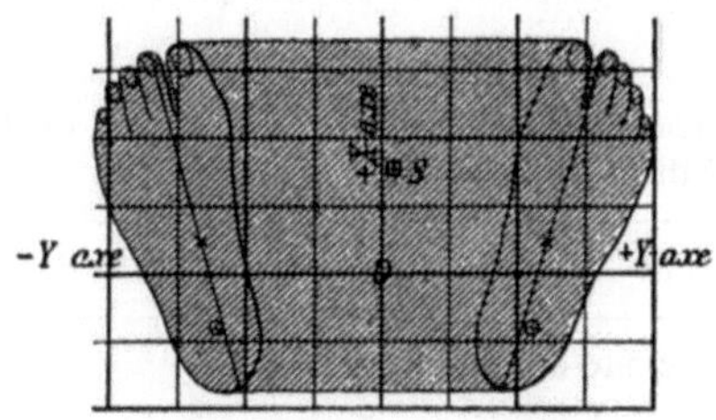

Fig. 30. Support area and its intersection ⊡ S by the line of gravity in the attitude of the subject holding his rifle forwards at arm's length, Figs. 14, 15. The origin O of the co-ordinates is the projection of the middle of the line connecting the centres of the hips. ⊙ Centre of the ankle joint, × centre of gravity of the foot. Scale: 1/10

The centre of gravity in this case lies 10 cm higher than the centres of the hip joints, approximately at the level of the middle of the fifth lumbar vertebra, 7.7 cm anterior to the centres of the hip joints. It lies outside the pelvis, even further anterior than the anterior border of the pelvis; it is 0.9 cm to the right of the median plane.

Figure 30 illustrates the intersection of the support area by the line of gravity. In this attitude, the line of gravity is quite far forwards in the support area but not so far as in the "present arms" attitude, as one might have expected. This results from the fact that here the subject displaces his body backwards at will, whereas in the "present arms" attitude the subject is compelled to incline the body forwards. However, the centre of gravity is displaced even further forwards, nearly 5 cm, than in the "present arms" attitude, such that it comes to lie completely outside the body. This makes the posterior inclination necessary. It is impossible to adopt the military attitude when the rifle is carried in this way.

It is also impossible to adopt the military attitude when a heavy load is carried in front of the body, as for example a bass drum.

To determine the influence exerted by the rifle held at arm's length on the position of the centre of gravity in this attitude of the body, we must calculate the co-ordinates of the centre of gravity in this attitude without the rifle:

$$x_0 = +\frac{78\,960}{58\,700} = +1.3$$

$$y_0 = +\frac{27\,195}{58\,700} = +0.4$$

$$z_0 = \frac{5\,477\,697}{58\,700} = 93.3.$$

Holding the rifle forward at arm's length in this attitude thus displaces the centre of gravity 6.4 cm forwards, 3.7 cm upwards, and 0.5 cm to the right.

Military Attitude at "Shoulder Arms" with Full Regulation Equipment, Figs. 16, 17

The co-ordinates of the centres of the joints are given in Table 44, the co-ordinates of the centres of gravity are given in Table 45. Using these co-ordinates and the weights of the pieces of equipment given in Table 33, we can calculate the co-ordinates x_0, y_0, z_0 of the total centre of gravity (Table 46).

Table 44. Co-ordinates of the centres of the joints in the military attitude at "shoulder arms" with full regulation equipment

Joint		x	y	z
Occipito-atlantal		+ 3	+ 1	151
Hip	r	0	+ 8.5	87
	l	0	− 8.5	87
Knee	r	− 5.5	+ 6	47
	l	− 5.5	− 6	47
Ankle	r	− 11	+ 4.5	6
	l	− 11	− 4.5	6
Posterior border of foot	r	− 15	+ 2	4
	l	− 15	− 2	4
Tip of foot	r	+ 9	+ 12	1.5
	l	+ 9	− 12	1.5
Shoulder	r	+ 3	+ 19	132
	l	+ 6.5	− 18	131
Elbow	r	− 1	+ 21	101
	l	+ 4	− 27.5	102
Wrist	r	+ 0.5	+ 24	74
	l	+ 26	− 18	91

Table 45. Co-ordinates of the centres of gravity for the military attitude at "shoulder arms" with full regulation equipment

		x	y	z
Head		+ 4	+ 1	155
Trunk		+ 1.2	+ 0.4	112
Thigh	r	− 2.4	+ 7.4	69.4
	l	− 2.4	− 7.4	69.4
Lower leg	r	− 7.8	+ 5.4	29.8
	l	− 7.8	− 5.4	29.8
Foot	r	− 4.7	+ 6.3	3
	l	− 4.7	− 6.3	3
Upper arm	r	+ 1.1	+ 19.9	117.4
	l	+ 5.3	− 22.5	117.4
Forearm	r	− 0.4	+ 22.3	89.7
	l	+ 13.2	− 23.5	97.4
Hand	r	+ 0.5	+ 23.5	69
	l	+ 29.5	− 15.5	90
Rifle		+ 6	− 11.5	143
Knapsack and back cartridge pouch		− 15.5	+ 0.4	114.5
Front cartridge pouch	r	+ 16	+ 12.5	95.5
	l	+ 16	− 12.5	95.5
Bayonet and spade		− 7.5	− 18	75
Haversack and water bottle		− 7.5	+ 18	75

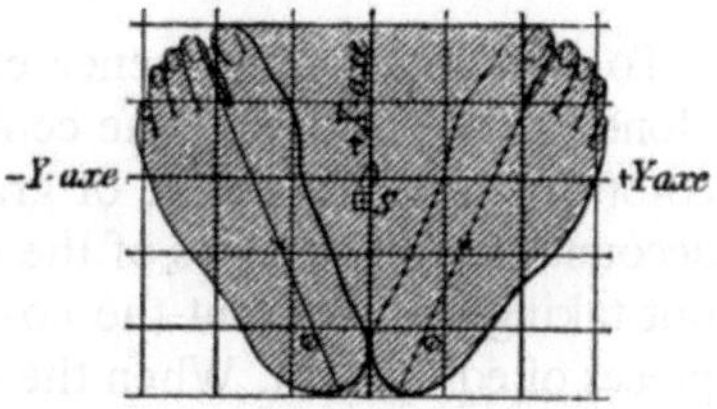

Fig. 31. Support area and its intersection ⊡ *S* by the line of gravity in the military attitude at "shoulder arms" with regulation equipment, Figs. 16, 17. The origin *O* of the co-ordinates is the projection of the middle of the line connecting the centres of the hips. ⊙ Centre of the ankle joint, × centre of gravity of the foot. Scale: 1/10

Consequently:

$$x_0 = -\frac{133903}{81970}, \quad y_0 = -\frac{37822}{81970}, \quad z_0 = \frac{7990167}{81970}$$

or: $x_0 = -1.6, \quad y_0 = -0.5, \quad z_0 = 97.5.$

The centre of gravity in this case lies 10.5 cm higher than the centres of the hip joints, approximately above the middle of the fifth lumbar vertebra, in the vertebra itself, 1.6 cm behind the centres of the hips. It lies 0.5 cm to the left of the median plane which results from carrying the rifle on the left shoulder.

Figure 31 represents the intersection of the support area by the line of gravity.

Table 46. Calculation of the co-ordinates x_0, y_0, z_0 of the total centre of gravity in the military attitude at "shoulder arms" with full regulation equipment

		$p_i x_i$	$p_i y_i$	$p_i z_i$
Head		+ 16 560	+ 4 140	641 700
Trunk		+ 30 072	+ 10 024	2 806 720
Thigh	r	− 16 320	+ 50 320	471 920
	l	− 16 320	− 50 320	471 920
Lower leg	r	− 24 102	+ 16 686	92 082
	l	− 24 102	− 16 686	92 082
Foot	r	− 4 935	+ 6 615	3 150
	l	− 4 935	− 6 615	3 150
Upper arm	r	+ 2 178	+ 39 402	232 452
	l	+ 10 494	− 44 550	232 452
Forearm	r	− 536	+ 29 882	120 198
	l	+ 17 688	− 31 490	130 516
Hand	r	+ 245	+ 11 515	33 810
	l	+ 14 455	− 7 595	44 100
Rifle		+ 28 200	− 54 050	672 100
Knapsack and back cartridge pouch		− 189 875	+ 4 900	1 402 625
Front cartridge pouch	r	+ 25 440	+ 19 875	151 845
	l	+ 25 440	− 19 875	151 845
Bayonet and spade		− 11 775	− 28 260	117 750
Haversack and water bottle (filled)		− 11 775	+ 28 260	117 750
Total		− 133 903	− 37 822	7 990 167

To determine the influence exerted by the rifle alone and by the equipment alone on the position of the centre of gravity in this case, we must calculate the co-ordinates of the centre of gravity in this attitude, firstly without taking into account the co-ordinates of the centre of gravity of the rifle, and secondly without taking into account the co-ordinates of the centres of gravity of the other pieces of equipment. When the centre of gravity of the rifle is neglected, the co-ordinates of the total centre of gravity are:

$$x_0 = +\frac{162103}{77279} = -2.1$$

$$y_0 = +\frac{16228}{77270} = +0.2$$

$$z_0 = \frac{7318067}{77270} = 94.7.$$

In this attitude, the living subject keeps his left elbow flexed and holds the left hand forward and a little medial, as is required for carrying the rifle. To determine how much this attitude displaces the centre of gravity forwards, upwards, and to the right, we assumed the position of the left arm to be symmetrical with that of the right. The co-ordinates of the total centre of gravity with full regulation equipment but without the rifle, with symmetrically held arms are:

$$x_0 = -\frac{202853}{77270} = -2.6$$

$$y_0 = +\frac{19064}{77270} = +0.2$$

$$z_0 = \frac{7297459}{77270} = 94.4.$$

The attitude of the left arm thus moves the centre of gravity 0.5 cm forwards and 0.3 cm upwards. The distance from the median plane remains unchanged: 0.2 cm. In this attitude, the centre of gravity does not lie exactly in the median plane, as might have been expected. This results from the fact that the subject inclined the upper part of the body a little to the right, as can be seen in Fig. 16.

If we neglect the centres of gravity of all the regulation equipment, including the rifle, the co-ordinates of the centre of gravity of the whole body in the attitude adopted as a consequence of the loading are:

$$x_0 = +\frac{442}{58700} = 0$$

$$y_0 = +\frac{11328}{58700} = +0.2$$

$$z_0 = \frac{5376252}{58700} = 91.6.$$

In this attitude of the body, when the left arm is symmetrical with the right, without the equipment and rifle, the co-ordinates of the total centre of gravity become:

$$x_0 = -\frac{40308}{58700} = -0.7$$

$$y_0 = +\frac{14164}{58700} = +0.2$$

$$z_0 = \frac{5355644}{58700} = 91.2.$$

To determine the influence exerted by the rifle alone in this attitude with the rifle on the shoulder, we simply need to omit the co-ordinates of the centres of gravity of the equipment (excluding the rifle). Thus:

$$x_0 = +\frac{28642}{63400} = +0.5$$

$$y_0 = -\frac{42722}{63400} = -0.7$$

$$z_0 = \frac{6048352}{63400} = 95.4.$$

To allow an easy comparison of the various results in the military attitude with full regulation equipment, the co-ordinates of the total centre of gravity are given in Table 47 for the different cases.

The regulation equipment thus displaces the centre of gravity relatively little. It is moved 3.2 cm upwards and 1.9 cm backwards, which appears when the figures for the symmetrical attitude of the body are compared with those for the military attitude. The radius of the femoral head reaches on average 2.5 cm.

Table 47. Summary of co-ordinates of the total centre of gravity for the military attitude with full regulation equipment

	In the military attitude[a]			In the symmetrical attitude		
	x_0	y_0	z_0	x_0	y_0	z_0
Body with full regulation equipment and rifle	– 1.6	– 0.5	97.5	–	–	–
Body with full regulation equipment without rifle	– 2.1	+ 0.2	94.7	– 2.6	+ 0.2	94.4
Body without regulation equipment, with rifle	+ 0.5	– 0.7	95.4	–	–	–
Body only	0	+ 0.2	91.6	– 0.7	+ 0.2	91.2

[a] Figures 16, 17

Table 48. Co-ordinates of the centres of the joints in the shooting attitude with full regulation equipment

Joint		x	y	z
Occipito-atlantal		+ 1	+ 7	147
Hip	r	– 5	+ 7	86
	l	+ 5	– 7	86
Knee	r	– 12	+ 9	46.5
	l	+ 5	– 14	46.5
Ankle	r	– 16	+ 15	6
	l	+ 2.5	– 21	6
Posterior border of foot	r	– 19.5	+ 11	4
	l	– 2.5	– 20.5	4
Tip of foot	r	0	+ 28	1.5
	l	+ 23.5	– 20.5	1.5
Shoulder	r	– 19	+ 14.5	135
	l	+ 7.5	– 12.5	133
Elbow	r	– 5	+ 36.5	128
	l	+ 36.5	– 1.5	120
Wrist	r	+ 16.5	+ 18.5	141
	l	+ 49.5	+ 11.5	141

Table 49. Co-ordinates of the centres of gravity in the shooting attitude with full regulation equipment

		x	y	z
Head		+ 1.5	+ 8.5	151
Trunk		+ 0.4	+ 2.7	109.8
Thigh	r	− 8.1	+ 7.9	68.6
	l	+ 5	− 10.1	68.6
Lower leg	r	− 13.7	+ 11.5	29.5
	l	+ 4	− 16.9	29.5
Foot	r	− 11.1	+ 18.3	3
	l	+ 8.7	− 20.5	3
Upper arm	r	− 12.4	+ 24.8	131.7
	l	+ 21.1	− 7.3	126.9
Forearm	r	+ 4	+ 28.9	133.5
	l	+ 42	+ 4	128.8
Hand	r	+ 19.5	+ 16	142.5
	l	+ 53	+ 12	142.5
Rifle		+ 47.5	+ 13	146
Knapsack and back cartridge pouch		− 18	− 8	114
Front cartridge pouch	r	− 2.5	+ 15	98
	l	+ 16	− 3.5	98
Bayonet and spade		+ 2	− 21	74
Haversack and water bottle		− 12	+ 12	74

The regulation equipment does not displace the line of gravity from the area of the hips, which indicates that the equipment was well designed. This calculation was verified by direct measurement on cadaver 3. After the cadaver had been frozen hard, lying supine, we equipped it as well as we could with the regulation equipment without the rifle and determined its centre of gravity in the way described above. We noticed a displacement of the centre of gravity 2.5 cm backwards and 3.5 cm upwards. The upwards displacement thus coincides to within 2 mm of our calculated result whereas the backwards displacement differs by 6 mm. This difference results from the fact that the frozen cadaver could not adopt the same attitude as the erect man whose back was somewhat arched forwards as a consequence of the knapsack.

As can be seen in Table 47, in the same attitude, the rifle on the left shoulder provokes a displacement of the centre of gravity 0.5 cm forwards, 3.8 cm upwards, and 0.9 cm to the left.

Shooting Attitude With Full Regulation Equipment, Figs. 18, 19

The co-ordinates of the centres of the joints are given in Table 48, the co-ordinates of the centres of gravity are given in Table 49. The calculation of the co-ordinates x_0, y_0, z_0 of the total centre of gravity is given in Table 50.

Table 50. Calculation of the co-ordinates x_0, y_0, z_0 of the total centre of gravity in the shooting attitude with full regulation equipment

		$p_i x_i$	$p_i y_i$	$p_i z_i$
Head		+ 6 210	+ 35 190	625 140
Trunk		+ 10 024	+ 67 662	2 751 588
Thigh	r	− 55 080	+ 53 720	466 480
	l	+ 34 000	− 68 680	466 480
Lower leg	r	− 42 333	+ 35 535	91 155
	l	+ 12 360	− 52 221	91 155
Foot	r	− 11 655	+ 19 215	3 150
	l	+ 9 135	− 21 525	3 150
Upper arm	r	− 24 552	+ 49 104	260 766
	l	+ 41 778	− 14 454	251 262
Forearm	r	+ 5 360	+ 38 726	178 890
	l	+ 56 280	+ 5 360	172 592
Hand	r	+ 9 555	+ 7 840	69 825
	l	+ 25 970	+ 5 880	69 825
Rifle		+223 250	+ 61 100	686 200
Knapsack and back cartridge pouch		−220 500	− 98 000	1 396 500
Front cartridge pouch	r	− 3 975	+ 23 850	155 820
	l	+ 25 440	− 5 565	155 820
Bayonet and spade		+ 3 140	− 32 970	116 180
Haversack and water bottle (filled)		− 18 840	+ 18 840	116 180
Total		+ 85 567	+ 128 607	8 128 158

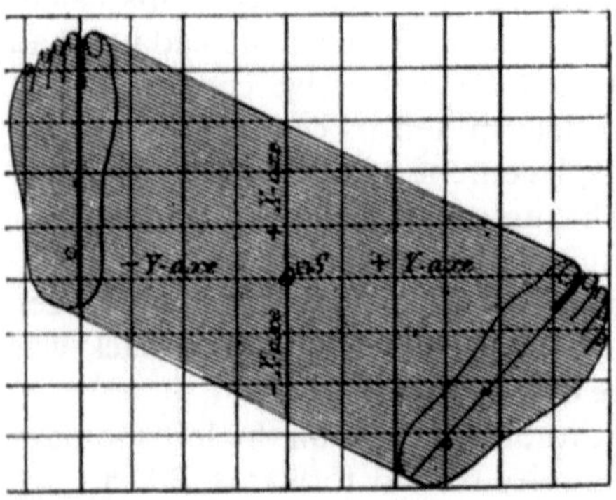

Fig. 32. Support area and its intersection ⊡ *S* by the line of gravity in the shooting attitude with regulation equipment, Figs. 18, 19. The origin *O* of the co-ordinates is the projection of the middle of the line connecting the centres of the hips. ⊙ Centre of the ankle joint, × centre of gravity of the foot. Scale: 1/10

Consequently:

$$x_0 = +\frac{85567}{81970}, \quad y_0 = +\frac{128607}{81970}, \quad z_0 = \frac{8128158}{81970}$$

or: $x_0 = +1.0, \quad y_0 = +1.6, \quad z_0 = 99.2.$

The centre of gravity in this case lies 13.2 cm higher than the centres of the hips, approximately at the level of the lower third of the fourth lumbar vertebra, at the anterior border of the vertebra. It lies a little more than 1 cm to the right of the median plane, which here cannot be determined precisely because of the rotation of the spine.

Figure 32 represents the intersection of the support area by the line of gravity.

To determine the influence exerted by the equipment on the position of the total centre of gravity in the shooting attitude, we calculated the co-ordinates x_0, y_0, z_0 excluding all the products related to the equipment except that related to the rifle. Thus:

$$x_0 = +\frac{300302}{63400} = +4.7$$

$$y_0 = +\frac{222452}{63400} = +3.5$$

$$z_0 = \frac{6187658}{63400} = 97.6.$$

The centre of gravity in the shooting attitude without the regulation equipment but with the same position of the trunk as when carrying the equipment lies at the level of the L_4-L_5 disc, about 3 cm in front of the disc, and more than 1 cm to the right of the median plane. It lies approximately 5 cm in front of the line connecting the centres of the two hips. The regulation equipment thus displaces the centre of gravity about 3.5 cm backwards and 1.6 cm upwards. In the shooting attitude without equipment (pp. 77–79), the centre of gravity was at about the same level but 7 cm in front of the line connecting the centres of the hips, and thus was 2 cm further forwards than in the attitude adopted when carrying the equipment, if this equipment is not taken into account. This demonstrates that the back was curved forward somewhat because of the equipment. This displaces the centre of gravity of the trunk backwards.

Since in the shooting attitude the centre of gravity was considerably displaced backwards by the equipment, the line of gravity was displaced backwards by the same distance. This results in a position of the line of gravity in the support area unfavourable for the stability against the recoil of the rifle. As we have seen for the shooting attitude without equipment, the stability of the attitude is increased when the line of gravity intersects the support area closer to its anterior than to its posterior border. To attain the same stability in the shooting attitude with equipment, the subject must lean forwards more than when not carrying the equipment.

Comparing Figs. 32 and 29, it appears that the line of gravity intersects the support area at the same place. This demonstrates that the man has spontaneously leaned his body forward.

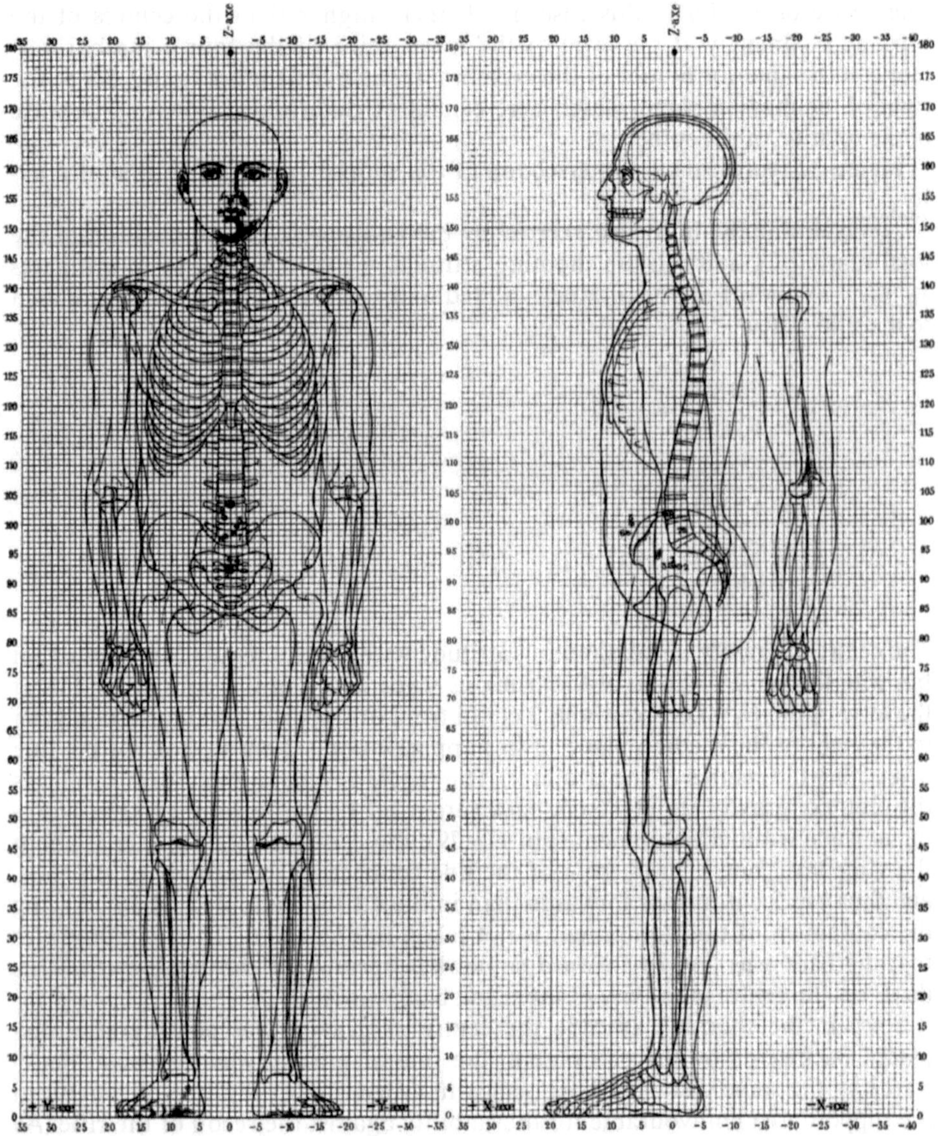

Fig. 33. Different positions of the centre of gravity of the whole body. *1.* Normal attitude, Fig. 6; *2.* comfortable attitude, Fig. 8; *3.* military attitude without regulation equipment, Fig. 9; *4.* "present arms" military attitude without regulation equipment, Figs. 10, 11; *5.* shooting attitude without regulation equipment, Figs. 12, 13; *6.* attitude with the rifle held at arm's length, Figs. 14, 15; *7.* military attitude with regulation equipment and rifle at "shoulder arms", Figs. 16, 17; *8.* shooting attitude with regulation equipment, Figs. 18, 19

Table 51. The co-ordinates x'_0, y'_0, z'_0 of the centre of gravity in relation to the normal attitude

	x'_0	y'_0	z'_0
– Normal attitude, Fig. 6	+0.2	0	92.7
– Comfortable attitude, Fig. 8	–0.8	0	92.3
– Military attitude without regulation equipment, Fig. 9	+0.4	0	92.7
– "Present arms" military attitude without regulation equipment, Figs. 10, 11	+2.9	+0.1	94
– Shooting attitude without regulation equipment, Figs. 12, 13	+7	–1	99.8
– Attitude with rifle held forward at arm's length (right), Figs. 14, 15	+7.7	+0.9	98
– Military attitude at "shoulder arms" with full regulation equipment, Figs. 16, 17	–1.6	–0.5	98.5
– Shooting attitude with full regulation equipment, Figs. 18, 19	+1.5	+1	101.2

Importance of the Position of the Line of Gravity in the Support Area

The position of the line of gravity in the support area gives a clue to the degree of stability of the different attitudes of the loaded and unloaded body that we considered. It does not, however, provide any information as to whether the position of the centre of gravity is favourable or unfavourable. The positions of the centre of gravity in the different attitudes are given in Fig. 33. In our living subject, the centres of the hip joints were 1 cm lower in most attitudes and 2 cm lower in the shooting attitudes than in the symmetrical form in Fig. 33. The Z co-ordinates thus have to be increased by 1 or 2 cm in order to be adapted to this diagram. Moreover, we must take into account that the line connecting the centres of the hip joints was no longer in the coronal plane in the shooting attitudes. This led to a correction of the co-ordinates X and Y. The projection of the feet gives an indication of the amount of rotation of the line connecting the centres of the hips. Considering these circumstances the co-ordinates x'_0, y'_0, z'_0 of the centre of gravity in relation to the normal attitude in Fig. 5 are given in Table 51.

The positions of the centre of gravity in Table 51 show directly what attitude the subject must adopt in order to bring the line of gravity as close as possible to the centre of the support area and in what positions of the centre of gravity the muscles must work the hardest, by arching the body, to achieve this posture. They also show which position is the closest to the normal attitude. Obviously, the centre of gravity lies in the most unfavourable position in the attitude with the rifle held forward at arm's length. The shooting attitude without regulation equipment seems to be nearly as unfavourable; the shooting attitude with equipment appears more favourable. In the shooting attitude without equipment, the centre of gravity lies outside the pelvis whereas it lies at the plumb line of the femoral heads in the shooting attitude with equipment. However, in the shooting attitude without equipment, the line of gravity falls almost in the centre of the support area, whereas in the shooting attitude with equipment a forwards inclination of the trunk is necessary to achieve this. In any event, the two shooting attitudes are unfavourable as far as the position of the height of the centre of gravity is concerned. Further practical conclusions can be drawn from Fig. 33.

On the Influence of the Flexibility of the Trunk on the Position of the Total Centre of Gravity

Because of the flexibility of the spine, the trunk is mobile and thus cannot be considered as a rigid mass. Therefore, it cannot be expected that, in the different attitudes, the centre of gravity of the trunk remains in the same place as that established in our frozen cadavers, i.e. in the plane through the centre of the occipito-atlantal joint and through the centres of the two hip joints. When the trunk is considerably bent posteriorly, the centre of gravity will remain at about the same height but it will be displaced backwards, and when the trunk is considerably bent forwards, its centre of gravity will be displaced anteriorly. This must be taken into account. In Figs. 6 and 9, the normal attitude of the living subject and the military attitude, we see that the centre of gravity of the trunk comes to lie in front of the aforementioned plane. Comparing these two figures with the attitude of the trunk in Fig. 5 shows, moreover, that the centre of gravity cannot lie as far forwards as in Fig. 5, because of the considerable backward movement of the trunk in Figs. 6 and 9. If, in Figs. 6 and 9, we assume the centre of gravity of the trunk to lie in the aforementioned plane, as we did in our calculation, we place this centre of gravity a little posterior to its actual position. If, on the other hand, in these figures we assume the centre of gravity of the trunk to lie inside the latter, at the place found by direct measurement, it is then somewhat anterior to its actual position. In Figs. 6 and 9 the centre of gravity is at its furthest from the plane through the occipito-atlantal and hip joints. We found on cadavers that the centre of gravity lies 2.5 cm from this plane in these cases. Thus, there exists a range of dispersal of 2.5 cm for the position of the centre of gravity of the trunk. A 1-cm displacement of the centre of gravity of the trunk forwards or backwards provokes a displacement of the centre of gravity of the whole body forwards or backwards by $\frac{25\,060}{58\,700} = 0.427$ cm for the unloaded body, by 0.395 cm for the body with the rifle, and by 0.306 cm for the body carrying the rifle and full regulation equipment. The maximum range of dispersal of 2.5 cm for the centre of gravity of the trunk in Figs. 6 and 9 thus corresponds to a range of 1.07 cm for the centre of gravity of the whole body. If we assume that the centre of gravity of the trunk lies in the middle of the area of dispersal, our co-ordinate X represents only a 0.5-cm discrepancy from the actual value. The total centre of gravity thus appears to lie 0.5 cm more anterior than we calculated. Our assumption that the centre of gravity of the trunk always lies in the plane through the occipito-atlantal and hip joints does not lead to significant errors in the attitudes of the body studied.

The average diameter of the femoral head of an adult man is 5 cm. Therefore, the centre of gravity of the trunk can be shifted anteriorly or posteriorly by up to about 2 cm without the coronal plane, which goes through the occipito-atlantal joint and centre of gravity of the trunk, leaving the area of the hip joints.

Effects of the Inclination of the Ground on the Attitude of the Body

What influence does the inclination of the ground exert on the attitude of the body? What influence does the regulation equipment exert on the posture of a subject on inclined ground? To answer these questions we used a 50-cm-wide and 2-m-long wooden board, which could be inclined progressively with a winch. The angle of inclination was read from a protractor.

When the board was inclined at 32°, which corresponds to very steep ground, the subject carrying regulation equipment could still stand but was unable to walk forwards up the steep surface. His body was considerably bent forwards. However, without the equipment and also by carrying the knapsack on his head and with the belt with the front cartridge pouches hanging from his neck, he could proceed even up this steep plane.

In this experiment, the subject had taken off his boots. Despite chalk being dusted onto the board and onto his socks, his feet slid. As this complicated the experiments, we carried out a second series in which sliding of the feet was hindered and the inclination of the plane was the only variable. A thin strip of wood was nailed to the board so that the soldier could stand with the heels of his boots against this strip, which hindered the sliding. Three young soldiers were the subjects for these experiments. Subject A with his regulation equipment could stand on the board up to an inclination of 41°30′ bending his body considerably forward, up to an inclination of 47° without equipment, and up to an inclination of 50° holding his knapsack in front of his chest. Subject B could stand up to an inclination of 41° with his regulation equipment, up to 47°–49° without his equipment, and up to 52° holding his equipment in front of his chest. Subject C could stand up to an inclination of 42° with his regulation equipment, up to 48° without his equipment, and up to 48° when holding his equipment in front of his chest.

It is not within our field of research to pursue this subject in further detail. It would be rather a matter for the army to carry out experiments on this problem and to decide to what degree regulation equipment impedes the displacement of the soldier on uneven ground.

The backwards displacement of the centre of gravity due to the equipment makes a forwards inclination of the trunk necessary in order to bring the centre of gravity above the centre of the support area and to ensure greater stability in the erect attitude. Similarly, when standing on an inclined surface, loading the lower back provokes changes in the attitude; the body bends forwards in order not to fall backwards. The steeper the plane and the more the centre of gravity is displaced backwards, the more the forwards bending increases. This bending can be carried out only in the ankles, hips, and spine. Individuals with stiff joints, usually in the older age range will be more impeded on mountainous ground than younger individuals with more mobile joints. Displacing the load upwards on the back facilitates the correction of the position of the centre of gravity by forwards bending of the body, since the bending capacity increases with the number of vertebrae involved.

It is due neither to chance nor fashion that in North Germany loads are often carried on the middle of the back, whereas in the more mountainous South Germany and in the Alps down to Italy they are carried on the head or as high as possible on the back. As far north as the river Main in Germany one can see women carrying baskets on their heads. In the Bernese Oberland, the heavy cheeses are carried as high as possible on the back. The girls of Capri carry heavy cases and bricks on their heads up the steep path to Anacapri.

On steep roads it is difficult to move with a heavy load and one must be more cautious in maintaining balance than on level ground. Loading the lower back increases the risk of falling backwards when walking on steep ground, when springing over a ditch towards a steep bank, or when walking over uneven ground.

Here we have simply mentioned a few further practical conclusions from our research on the centre of gravity without, however, wishing to enter in detail an area of study in which we are not competent.

Subject Index

MIX
Papier aus verantwortungsvollen Quellen
Paper from responsible sources
FSC® C105338

If you have any concerns about our products,
you can contact us on
ProductSafety@springernature.com

In case Publisher is established outside the EU,
the EU authorized representative is:
Springer Nature Customer Service Center GmbH
Europaplatz 3, 69115 Heidelberg, Germany

Printed by Libri Plureos GmbH
in Hamburg, Germany